THE CONTROL OF
OIL

THE CONTROL OF
OIL

EAST-WEST RIVALRY
IN THE PERSIAN GULF

Alawi D. Kayal

Routledge
Taylor & Francis Group

LONDON AND NEW YORK

First published in 2002 by
Kegan Paul International

This edition first published in 2009 by
Routledge
2 Park Square, Milton Park, Abingdon, Oxfordshire OX14 4RN

Simultaneously published in the USA and Canada
by Routledge
711 Third Avenue, New York, NY 10017

First issued in paperback 2016

Routledge is an imprint of the Taylor & Francis Group, an informa business

British Library Cataloguing in Publication Data
A catalogue record for this book is available from the British Library

ISBN 13: 978-1-138-96668-0 (pbk)
ISBN 13: 978-0-7103-0768-2 (hbk)

Publisher's Note
The publisher has gone to great lengths to ensure the quality of this reprint
but points out that some imperfections in the original copies may be
apparent. The publisher has made every effort to contact original copyright
holders and would welcome correspondence from those they have been
unable to trace.

Contents

List of Tables

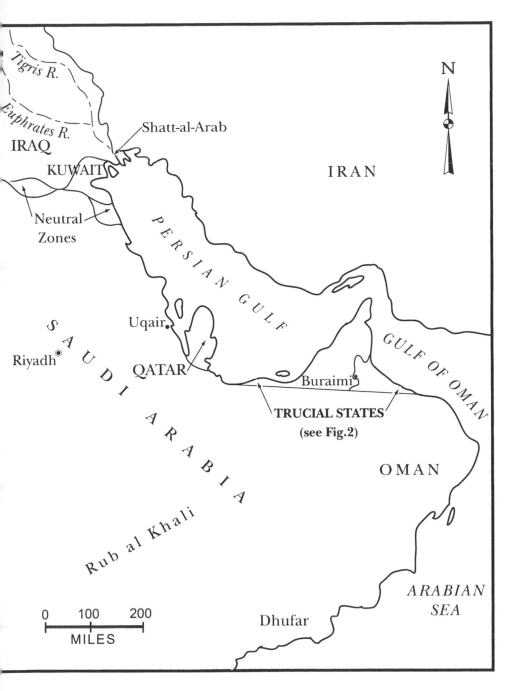

Fig. 1. Persian Gulf Area
Compiled from various sources.

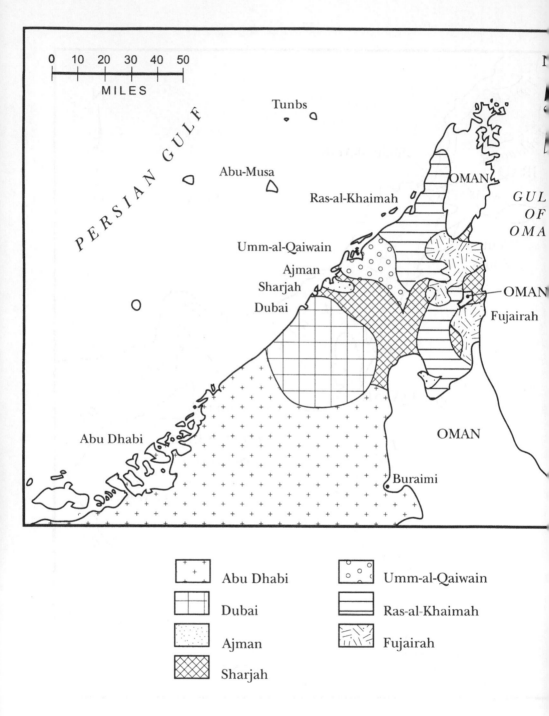

Fig. 2. Omani Coast (Trucial States)
Compiled from various sources

Preface

Today, the United States stands at the helm of the Middle East. Her control of the Persian Gulf oil is masterful and firm. The world's sole superpower's control of this area is a factor in the exercise of her world leadership.

Oil is of strategic significance. It is almost unique in this respect. Its significance lies in its permeating nearly every aspect of the economic life of present-day nations. It consumes governments and precipitates wars. The world's developed economies are heavily dependent on oil, and no reasonable substitute for it is anticipated in the foreseeable future.

The bulk of the earth's known oil reserves, more than 70 percent, is concentrated in the Persian Gulf area. And although alternative energy sources have been vigorously pursued, the United States continues, since 1970, to import from the Persian Gulf 24 percent of needed oil for her own consumption. Japan and Europe, of course, are in varying degrees totally dependent on oil imported from this area. The region, therefore, will continue to be the scene of a control challenge amongst the powers of the world. For the control of this area and its individual states presses the oil-consuming countries into accommodation to the directives of the controller.

With such magnitude of dependence on oil from the Gulf area, and with the 1973 oil shock vivid in the minds of the US security planners, the Iranian uprising of 1979 and its repercussions alerted President Jimmy Carter in January 1980 to reassert the United States policy in the Gulf in what came to be known as the Carter Doctrine:

> Let our position be absolutely clear. An attempt by any
> outside force to gain control of the Persian Gulf region will be
> regarded as an assault on *the vital interests of the United States*

of America, and such an assault will be repelled by any means necessary, including military force. (Emphasis added.)[1]

The vital interests of the United States cannot, therefore, be left to the complexities of Arab politics and Arab–Israeli conflicts. Consequently, the United States, a military and economic might, has embarked upon a path of finding opportunities in crises in the Middle East to strengthen her influence and affirm her presence there. Piece-by-piece she has worked out ways to formulate policies aimed at insulating oil from Middle East politics – and world politics for that matter.

Since this study was completed thirty years ago there have been several major events related to the control of the flow of Gulf oil: the 1973 Arab–Israeli war and the ensuing oil embargo; the Iranian revolution and the fall of the Shah; the Iran–Iraq war; the participation in and nationalization of the ownership of oil, and the attendant cancellation of the concession agreements; the diminution of the powers of the major oil companies; the increased strength of OPEC and the sustained hike in the price of oil; the break-up of the Soviet Union which altered Russia's strategic position as a superpower; the invasion of Kuwait by Iraq; the Gulf war; the sanctions imposed on Iraq and Iraq's subsequent isolation. Casting a long shadow over all these events is the Palestinian plight with its resultant continuing Arab–Israeli conflict.

Oil in the Middle East is very much intertwined with the Arab–Israeli struggle. The path followed by the United States in formulating her policies is of seemingly contradictory aims. It ranges from making use of Arab fragmentation, through beefing up Israel's military capabilities and guaranteeing its arms supremacy, to the direct presence of United States military forces in the area.

In the process of these policies Egypt, the heart of the Arab nation, was alienated at Camp David in 1979. Soon she found herself expelled from the Arab League, with the effect of segregating her from the Arab–Israeli conflict and Arab politics, and thereby the politics of oil. Egypt's expulsion carved a deep wound in the Arab heart. When she finally returned she had been away for so long that the wound remains wide open and bleeding in what we see of the Arab disarray, and in what is happening to the Palestinians and the Palestinian cause.

The Palestinian tragedy is the core of the Arab rage. One needs to remember that the creation of Israel called for the establishment of an exclusively Jewish state in a country that was predominantly Arab. An entire society of Palestinian Arabs has been uprooted from its homeland and thrown into refugee camps in order for its land to house people brought in from all over the world – from Eastern Europe, from Argentina, from South Africa, from all corners of the earth. The name of a country called 'Palestine' was removed from the world's atlases and the name of a country called Israel was substituted.[2] It was that simple. And what was left of Old Palestine – the West Bank and the Gaza Strip – is now in siege under the most repressive measures and humiliations.

Myth and sophistry aside, all of the Palestinian sufferings were committed under United States auspices and with her full support. Since 1972 the United Nations Security Council has voted on more than 30 resolutions condemning the atrocities committed by Israel against the Palestinians. All 30 of these resolutions were vetoed by the United States, causing them to fail. In most of these cases, the United States stands out as the sole dissenter against the other 14 members of the Security Council, who voted in support of the resolutions.[3] Too, America's ability to arbitrarily decide which United Nations resolutions she will press for adherence (Iraqi sanctions) and which she will flagrantly disregard (Israel's withdrawal from the territories occupied as a result of the 1967 war), is causing deep anger and bitterness among the Arab people. People everywhere can neither accept nor tolerate the innate injustice of this double standard, especially when it is practiced by a country which is in the position of leadership of this world, with the responsibility of adjudicating the rule of law and the declared principles of democracy and human rights.

It is in this environment one should view the Arab revulsion and therefore the nature of Middle East politics and the politics of oil. It must be said, however, that the Arabs do not have a special appetite for hurting Americans or American interests in the Gulf area, nor in the Arab world. It is the unqualified, unconditional, extensive support of the State of Israel, a circumstance that amazes so many Arabs, and so many Americans, too, which is the real source of the rift.

The Iraqi invasion of Kuwait further intensified the fragmentation amongst the Arab countries. This time the rupture touched on a psychological phenomenon difficult to reverse – mistrust. Nevertheless, the ejection of Iraq from Kuwait necessitated the build-up of the US military might in the Gulf, a build-up which seems to be taking the course of being permanent. As of now, the United States is busy devising a missile defense system she says will check the rising power of what she calls the 'rogue' states – Iran, Iraq, Libya. These states, especially Iran, are in possession of medium-range ballistic missiles. The spread of these missiles is altering the strategic balance in the Middle East, placing more targets in jeopardy, including the United States' extensive military presence in the area. The US is now there on land, at sea, in the air, and even in space.

And the reason the United States is there is to protect the flow of oil. There is no other reason. Over the years, the United States has set about formulating policies which bar the Arabs from using oil as a leverage in attaining their objective. Oil, or the flow of oil, is wholly insulated from both Arab and world politics, politics which make dependent states beholden to the controller.

Has oil then been relegated to the level of being insignificant? No, it has not. Oil is still in the heart of present-day nations. It is the base of our civilization, as Daniel Yergin, the author of *The Prize* described it.[4] But if you ask me, 'How is it manipulated?' my response would be, 'Who is in control?'

This work narrates the history of the world's power struggle over the control of oil in the Persian Gulf from the time of the signing of the earliest oil concessions in 1901 until 1971. The first section outlines the social and political environment in the Persian Gulf area insofar as it throws light on the nature of the power struggle that preceded the granting of the first concession. The second part extends from the events of World War II to the rise of the Organization of Petroleum Exporting Countries (OPEC) in 1960. The final part examines OPEC as a power center.

This discussion is about the politics of oil, and therefore needed the support of men of principles at the University of Colorado: Professors Edward J. Rozek and Professor Dayton McKean. To both men I acknowledge my debt and gratitude. I also owe much to Professor Ragaei El Mallakh who provided useful information on the economy of oil in the Gulf.

I am indebted to Abdulaziz Bagdady, Emily Cooper, Claire Sayre, my brother Mohideen Kayal, and Abdulaziz Addughaither, who assisted me in preparing the study. I am grateful to my government for sponsoring my education and making it possible for this study to materialize. Finally, the work would have been harder had it not been for the patience and support of my wife.

Professor Zuhayr Mikdashi of Université de Lausanne was invaluable in getting this publishing project off the ground. Peter Hopkins of Kegan Paul managed to remain unflappable, spread encouragement and perspective, and did his best to keep us all on schedule. Not an easy task. Matthew Garcia, Ali Mostashari, Sharon Morris, Lee Bowers, Bette James and Laura Nunn performed diligent and tedious work. To them all I owe much appreciation.

Last, but not least, Janice Murray persevered through the ups and downs, and had the heart to carry on.

This work would have been harder had it not been for the enduring patience and support of my wife, Huda. My son Amro's unwaivering encouragement and belief in this publication, my daughter Hawazen's confidence in me and my work, and my son Hesham's faith in my judgment, together with their love and trust, have carried me far.

Alawi D. Kayal
August 2001

Endnotes

[1] Warren Christopher, Harold H. Saunders, Gary Sick and Paul Kriesberg, *American Hostages in Iran: The Conduct of A Crisis*. New Haven: Yale University Press, 1985, 112.

[2] J.G. Bartholomew, *The Times Survey Atlas of the World*. London: The Times, 1922.

[3] US Department of State, '1972-1994 Voting Practices in the United Nations Bureau of International Organization Affairs.' Report to Congress. Submitted Pursuant to Public Law 101-167. March 31, 1995. 'Lessons to be Learned from 66 UN Resolutions Israel Ignores.' Washington DC: Washington Report on Middle East Affairs, 1993.

[4] Daniel Vergin, *The Prize*. New York: Touchstone, 1991.

1

The Demand for Oil

Behind the power struggle for the control of world oil reserves is to be found an ever-rising demand to attempt to satisfy the insatiable energy-hunger of the industrialized world. The figures are dramatic. During the 1960s a remarkable growth average of about seven percent a year in demand was consistently maintained. World oil demand has, then, doubled, and oil production gone up from a daily average of 21 million barrels per day (bpd) in 1959 to 42 million bpd in 1969. During that period, 100 billion barrels were consumed. On the basis of a very conservative growth rate of five percent, world oil production is expected to average 105 million bpd in 1989. By then, aggregate requirements would reach a level of 532.9 billion barrels – just about the world proven oil reserves now in existence.[1]

This chapter gives a detailed analysis of the oil requirements of the industrialized countries. It will also deal with these countries' oil supply positions so as to throw some light on the nature of the power struggle behind the world of oil.

United States

In 1970, oil provided the United States with 43.7 percent of her energy requirements. Oil was then followed by natural gas, 32.3 percent; coal, 20 percent; hydroelectric, 3.8 percent; and nuclear power 0.6 percent (see Table 1). In 1970, United States oil consumption averaged 14.8 million bpd. This is 672,000, or 4.9 percent, above the 1969 daily average consumption.[2] The aggregate consumption for 1970 reached 5.4 billion barrels – a record consumption which, according to the British Petroleum Press Service, is the 'equivalent of 1,100 gallons of oil products for every man, woman and child in the US.'[3] This consumption is comprised of fuel oil, which rose in 1970 over the preceding year by 14.6 percent to 2.3 million bpd; gasoline, which rose by 4.7 percent to 5.8 million bpd; distillate fuel oil, which rose by 2.7 percent to 2.5 million bpd; kerosine-type fuel, which rose by 5.5 percent to 0.7 million bpd. Naphtha jet oil, however, declined by 18.9 percent to 0.2 million bpd, as did aviation gasoline by 24.3 percent to 0.05 million bpd.[4]

TABLE 1
Energy Forms: Percentage Shares of Consumption and Projection Rate, United States

	1965 %	1970 %	1975 %	1980 %
Oil	43.1	43.3	43.7	43.3
Gas	29.9	32.3	31.1	29.1
Coal	23.1	20.0	17.2	14.7
Hydroelectric	3.9	3.8	3.4	2.9
Nuclear	-.-	0.6	4.6	10.0
Total	100.0	100.0	100.0	100.0

Source: 'US Dependence on Oil Imports,' Petroleum Press Service, XXXVIII, No. 5 (May, 1971), Table 1, 170.

On the supply side, total US domestic production of liquid hydrocarbons amounted to 11.3 million bpd in 1970, an increase of 4.9 percent over the year before. Production of crude oil rose in 1970 by 4.8 percent to 9.1 million bpd, natural gas liquids rose by 5.8 percent to 1.7 million bpd, lease condensate rose by 1.1 percent to 0.5 million bpd, and other liquid hydrocarbons amounted to 0.02 bpd.[5]

The difference between domestic oil production and consumption was made up for by imports, which in 1970 amounted to 1.3 million bpd of crude and 2.1 million bpd of products. Imports represented 23 percent of the total United States domestic petroleum consumption.[6]

The gap between oil supply and demand has been the cause of a deep concern among United States officials. The gap, however, appears to be a basic deficiency in conventional petroleum reserves and, according to Monty Hoyt, 'few in the oil industry think the day will be regained when US will be self-sufficient in its production of crude oil.'[7] In 1969, United States proven oil reserves amounted to 29.6 billion barrels. This is an accelerated declining rate of 3.5 percent from the year before. It is greater than either the 2.1 percent decline of 1968 or the 0.2 percent decline of 1967.[8] United States oil reserves represent only 8.2 percent of the free world (excluding the Communist bloc) oil reserves and they are just enough for 9.5 years at the 1970 rate of production.[9]

The dwindling of United States oil reserves is further intensified by the decline in the number of oil wells drilled. The number of wells drilled in 1970 dropped to 29,540, or 8.5 percent of the 1969 level. Exploratory well drillings (wildcat wells) declined by an even greater rate – 13 percent. They dropped from 13,736 wells in 1969 to 11,650 in 1970. This is a continuation of a fourteen-year decline of 48 percent of the 1956 level.[10]

United States domestic oil reserves are further exhausted by reasons of ecological requirements that forced coal out of the market, the shortage in natural gas production, and the technical problems facing the development of nuclear energy. In 1951, Professor Halford L. Hoskins made the prediction that: 'Atomic energy may partially supersede oil some day as a driving force in the modern state, but not in the visible future.'[11] Twenty years have passed since that prediction was made, and nuclear energy only constitutes 0.6 percent of the United States total energy requirements (see Table 1).

Thus after years of energy abundance, the United States finds itself, to use the words of John D. Emerson, the Chase Manhattan Bank's energy economist, facing the worst energy shortage of the twentieth century.[12] One might expect a breathing spell from the development of the Alaskan North Slope and shale oil, the gasification of coal and the development of nuclear energy. The development of these scarcer and scarcer resources, however, takes time, and time, according to Gene Morrell, US Deputy

Assistant-Secretary for Interior, is the one thing which the United States does not have. These resources require sufficient funds to explore extensively and to drill deep enough to unlock them.[13] It is therefore believed that, in all probability, oil will continue to be the dominant feature in the United States energy pattern, and no one is seriously expecting other energy substitutes to take its place.[14] Accordingly, Morrell, looking at the United States supply position, had this to say:

> Our demands for petroleum are so tremendous and growing so rapidly, and the state of our domestic capacity to match those demands so impaired, that no matter what we do imports are likely to increase their share of our total oil supply over the foreseeable future.[15]

For a policy that wishes to maintain self-sufficiency in energy requirements, this reality must be the source of great anxiety. Nevertheless, a study put out by the Department of Interior projected domestic production of liquid hydrocarbons to amount to 13.1 million bpd in 1975 – an increase of 16 percent of the 1970 level. The study estimated that in 1975 imports would reach 2.5 million bpd. Canada, according to the US Interior Department, would then provide the United States with 1.5 million bpd, Mexico with 0.05 million bpd, Venezuela with 0.50 million bpd, and the Eastern Hemisphere with 0.45 million bpd.[16] The Interior Department then has tried to keep the amounts of imports, particularly those from the Eastern Hemisphere, at a minimum.

There are those, however, who believe that the United States is vastly underestimating its future energy requirements. Dr Paul McCracken, Chairman of the President's Council of Economic Advisers, sees, since 1966, the emergence of a trend of energy consumption rising at a rate faster than that of economic growth. According to E.R. Heydinger, Manager of Marathon Oil Company's economic division, this trend is attributable to 'a definite broad-based shift in national social values toward a more relaxed life style but without sacrifice of material affluence.'[17]

4

With this tendency in mind, it is not surprising to see, according to a study by Chase's John Emerson that the United States energy consumption will hit, in 1980, 52 million bpd in oil equivalent and that half of the United States oil requirements will be imported.[18]

The study indicated that of the 52 million bpd of oil equivalent required in 1980, oil contribution would be around 21.5 million bpd.[19] Natural gas followed with a contribution of 16.5 million bpd. The rest would be provided for by coal, hydro, and nuclear power. Emerson however, is of the opinion that no more than 11 million bpd of natural gas would be available from all US sources. Accordingly, the study assigned to oil an additional demand of 3.5 million bpd and to coal the equivalent of two million bpd of the potential natural gas consumption in 1980. Hence, the demand for oil would by 1980 reach the level of 25 million bpd.

On the supply side, the study indicated that by 1980 domestic oil production would provide the country with 12.5 million bpd – exactly half the amount required. Imports would then make up for the other half. Canada would be able to supply the United States with a maximum of 2.0 million bpd, other Western Hemisphere countries with 3.2 million bpd, and the Far East with a token amount of 0.2 million bpd. An amount of 7.1 million bpd – more than 50 percent of the total oil imports – would be left to be filled from sources in the Persian Gulf area and Africa.

Despite the fact that the United States controls over 60 percent of the oil reserves of the Persian Gulf area, the United States, nevertheless, followed a policy of curbing its imports from that part of the world. However, the preceding analysis indicates that the United States is passing from a period of energy abundance, and soon she will be rivaling other powers in the international oil market for the direct use of oil. At this point, it is worthwhile to note that oil is not sought and controlled by the superpowers of the world solely for the satisfaction of home demands; it is a valuable investment, as well as a strategic commodity.

Soviet Union

The Soviet Union is the second largest oil producing and consuming country in the world. Until 1960, coal was the major source of energy consumed in the Soviet Union. At that time coal formed 56.2 percent of the total Soviet Union energy requirements; followed by oil, 26.1 percent; and natural gas, 8.4 percent (see Table 2). The trend, however, is obviously in favor of the two latter forms of energy. In 1965, coal went down by almost 20 percent to 45.3 of the total energy requirements of the USSR. At the same time oil went up by 13 percent to 29.5 percent and natural gas by 104 percent to 17.2 percent, thus making a combined share of 46.7 percent of the Soviet Union energy consumption in 1965. The consumption of coal further dropped in 1969 to 40.7 percent while that of oil rose to 31.7 percent and natural gas to 20.7 percent. Projected energy consumption in 1975 was put at 33.7 percent for coal, 35 percent for oil and 24.7 percent for natural gas. In 1980, the projection is estimated at 30 percent, 36.5 percent and 24.9 percent for coal, oil, and natural gas, respectively.

TABLE 2
Energy Forms: Percentage Share of Consumption and Projection Rate, Soviet Union[1]

	1960 %	1965 %	1969 %	1975 %	1980 %
Coal	56.2	45.3	40.7	33.7	30.0
Oil	26.1	29.5	31.7	35.0	36.5
Natural gas	8.4	17.2	20.7	24.7	26.9
Miscellaneous	9.3	8.0	6.9	6.6	6.6
Total	100.0	100.0	100.0	100.0	100.0

[1]*Calculated from: US Department of Interior, Bureau of Mines, 'The Mineral Industry of the USSR,'* Minerals Yearbook, Area Reports: International, *Vol. IV (Washington DC: US Government Printing Office, 1971), Table 7, 749.*

In 1969, actual oil consumption amounted to 202.9 million tons per year, or 4,058,000 bpd, an increase of 6.7 percent over 1968 consumption. Oil consumption rose in 1970 by 7.0 percent to 217 million tons per year, or 4,340,000 bpd. The US Bureau of Mines projected Soviet oil consumption levels of 276 and 342 million tons in 1975 and 1980, respectively (Table 3).

On the supply side, Soviet oil production in 1969 reached 328 million tons, or 6,560,000 bpd – an increase of six percent over a year before. In 1970 production went up by seven percent to 350 million tons, or 7,000,000 bpd. Soviet production is expected to reach 445 million tons per year in 1975 and 540 million tons in 1980 (see Table 3).

This position provided the Soviet Union with a handsome surplus of crude oil and products to export to Eastern and Western markets. Soviet exports went up from 64.4 million tons in 1965 to 90.8 million tons in 1969. Crude oil and petroleum product exports were put at 97 million tons in 1970. The levels for 1975 and 1980 were put at 129 million and 150 million tons, respectively.[20]

Signs of Soviet difficulty in coping with the rising oil demands were, however, depicted in a number of Soviet behaviors and measures. They were seen in the slowdown of oil exploration, in the leveling out of exports, and in the conclusion of oil agreements with a number of foreign countries for providing the Soviets with foreign oil. It appears, therefore, that in the next few years the Soviets may find it difficult to produce enough oil to satisfy their own needs as well as those of Eastern Europe.[21]

In 1969, Soviet proven oil reserves were put at 3.6 billion tons, or 26.3 billion barrels, a declining production/reserve ratio, according to the US Bureau of Mines. Therefore proven oil reserves were estimated to be just enough for 14-15 times the 1968 Soviet production.[22]

TABLE 3

Petroleum Supply And Demand of the USSR
(Million Metric Tons)

Item	Actual					Planned and Estimated	
	1966	1967	1968	1969	1970	1975	1980
Crude oil							
Domestic	265.1	288.1	309.2	328.0	350.0	445.0	540.0
Imports		(a)				5.0	10.0
Exports							
To communist countries	25.5	27.2	32.0	38.1	42.0	62.0	80.0
To non-comm. countries	24.8	26.9	27.2	25.8	27.0	35.0	40.0
Total	50.3	54.1	59.2	63.9	69.0	97.0	120.0
Crude to refineries	214.8	234.0	250.0	264.1	281.0	353.0	430.0
Refined oil							
Output from crude	185.0	199.0	121.0	224.0	239.0	300.0	366.0
Natural gas liquids	3.0	3.7	4.1	4.7	5.0	8.0	12.0
Imports	1.7	1.4	1.0	1.1	1.0	(a)	(a)
Exports							
To communist countries	6.7	8.0	9.5	9.5	10.0	11.0	12.0
To non-comm. countries	16.6	16.7	17.5	17.4	18.0	21.0	24.0
Total	23.3	24.7	27.0	26.9	28.0	32.0	36.0
Apparent consumption	166.4	179.4	190.1	202.9	217.0	276.0	342.0

(a) Insignificance

Source: US Department of Interior, Bureau of Mines, 'The Mineral Industry of the USSR,' Minerals Yearbook, Area Reports: International, Vol. IV (Washington DC: Government Printing Office, 1971), Table 12, 758.

To be sure, potential oil reserves in the Soviet Union are large enough to sustain higher production. But as the most shallow and more accessible structures are exhausted, attention gets drawn to remote and uninviting areas. The Soviets have put most of their recent exploratory efforts in Siberia, which, like the Alaskan North Slope, is fraught with difficulties: of drilling deeper, of the more exploratory wells, of increasingly complex geological conditions, and of remoteness of drilling location – all of which take more time and require huge capital expenditure with the obvious adverse impact on cost.[23]

As a result of these geological difficulties, the effectiveness of exploration in the country as a whole, according to *Neftyanage Khozaistro*, a Russian oil industry journal, is not increasing, and in a number of areas it is diminishing.[24] The year 1969 witnessed, for the first time since the Soviets resumed their oil exportation early in the 1950s, a decline in the rate of Soviet exports. Soviet oil exports then dropped by seven percent to 939,000 bpd from the level of 1,008,000 bpd attained in 1968. Soviet oil imports, on the other hand, trebled in 1969, averaging 53,000 bpd.[25] The year 1969 also marked the beginning of Soviet oil consumption increasing at a rate faster than their production.[26]

These are just signs of how the Soviets are finding it difficult to satisfy all their oil requirements from areas within their borders. In 1971 Soviet planners had to scale down the rate of oil production to less than six percent, thus realizing an increment of only 400,000 bpd. Of these 400,000 bpd, 260,000 bpd, or 65 percent, were to be brought from costly areas, i.e., Western Siberia.[27]

Indeed, the US Bureau of Mines predicted a Soviet oil deficit to develop in 1975 and 1980 of five million and 10 million tons, respectively (Table 3). These figures, however, were challenged by Rafkhat Mingareev, deputy head of the USSR Ministry for Oil Extraction. Mingareev predicted a level of oil consumption for the whole Soviet bloc in 1980 of 680 million tons, or 13.6 million bpd. By then, he predicted production would reach the level of somewhere around 625-645 million tons.[28]

If allowance is made for the Eastern Europe production of 25 million tons, there would develop a probable deficit of 10-30 million tons by 1980. These estimates, however, do not take into consideration Soviet insistence on maintaining their oil exports to Western markets. Also, the estimates appeared to refer to refined products, so that in terms of crude oil they would be substantially higher. Accordingly, a deficit of far greater volume than the 10 million-30 million tons could reasonably be expected.[29]

In fact, the London *Economist* predicted a deficit of 100 million tons (2 million bpd) and possibly 150 million tons (3 million bpd) by 1980.[30] The *Economist* first considered the target set by Soviet planners for the production of 460 million tons of oil in 1975. It then made the observation that in the past decade Soviet oil output increased by a steady 20 million tons a year and that there is no reason to see acceleration at a time when Russian economic growth is slowing down. Thus, 560 million tons was given as a realistic figure for Soviet oil production in 1980. By then Soviet oil consumption would reach the annual rate of 613 million tons, creating a possible deficit of around 50 million tons.

This is not the whole story, however. In 1980, the countries of Eastern Europe would be consuming oil in the range of 140 million-170 million tons, of which they themselves would be producing around 30 million. A deficit is thus foreseen, the bulk of which – say 100 million tons – the Soviets would insist on providing.

The above analysis hinted at the possibility of the Soviet Union finding it difficult to satisfy their oil requirements and those of the Soviet bloc from areas within their domain. Clearly the gap could be filled most conveniently from places just across the border, i.e., the Persian Gulf area. Accordingly, a number of steps were taken in this direction. The Soviets concluded an import agreement with Iran for what is termed the 'largest imports in Soviet history;'[31] another with Iraq for the development of the North Rumaila oil field in exchange for Iraqi oil; a third with Kuwait; a fourth with Syria; and a fifth and sixth with Libya and Algeria. Details of these agreements will be reserved for a later chapter. The point to be emphasized here is that when V.D. Shashin, Minister of Petroleum Industry of the Soviet Union, denied the possibility of Russia being turned into a net oil importer, the *Oil and Gas Journal* hurried to point out that Shashin 'failed to clarify how recent Soviet oil – aid deals in the Middle East and North Africa – some of them involving Russians taking crude oil in payment—fit into this picture.'[32] Incidentally, satisfaction of home energy requirements cannot be construed as the only justification for a superpower to seek control over a portion of the oil reserves of the world.

Western Europe

All over Western Europe, the demand for petroleum products expanded even more sharply than that projected by the forecasters. In 1970 the area's over-all increase in inland sales of petroleum products exceeded the rate of 11 percent recorded in 1968 and 1969.[33]

In 1970, inland energy consumption of the European Economic Community amounted to 847.8 million metric tons of coal equivalent – an increase of 70 million tons over the preceding year. Of the total energy consumed in 1970, oil consumption amounted to 496.2 million metric tons of coal equivalent, or 58.6 percent. Oil then increased by 57.3 million tons of coal equivalent. Thus, while total energy consumption increased by 9.5 percent in 1970 over the preceding year, oil rose by 13.1 percent. Total energy consumption in 1971 was put at 894.2 million metric tons of coal equivalent – an increase of 5.5 percent. Oil consumption was then estimated at 536.2 million tons of coal equivalent, an increase of 8.1 percent.[34] Oil represented in 1971 60 percent of the total inland energy consumption of the European Common Market (Table 4).

In the United Kingdom, total demand for primary energy in 1960 amounted to nearly 265 million tons of coal equivalent. Then coal provided the country with 196.7 million tons, or 74.3 percent; oil with 65.5 million tons of coal equivalent, or 24.7 percent; and nuclear energy and hydroelectricity with around one percent. In 1970 total inland consumption of energy reached 330 million tons of coal equivalent – an increase of 24.5 percent over 1960. During this period, coal went down by 20.2 percent, making a contribution to the energy supply of 47.3 percent or, in tonnage terms, 156 million tons. Oil, on the other hand, rose by 121.3 percent for a share in the total energy consumption of 43.9 percent, or 145 million tons of coal equivalent. Natural gas contribution was put at 4.9 percent with nuclear energy and hydroelectricity estimated at 3.9 percent (see Table 5). Although the figures forecast for coal and oil consumption in 1975 – put at 118 and 146 million tons of coal equivalent, respectively – were already reached in 1970, the forecast, nevertheless, suggests in 1975 contribution share for coal of 33.7 percent; oil, 41.7 percent natural gas, 14.0 percent; and nuclear energy and hydroelectricity, 10.6 percent.[35]

TABLE 4
Energy Forms: Percentage Share of Inland Consumption
European Economic Community[1]

	1969 %	Provisional Forecast 1970 %	1971 %
Coal	26.4	23.1	20.5
Lignite	4.2	4.0	4.0
Oil	56.4	58.6	60.0
Natural gas	7.0	8.5	9.8
Primary electricity	5.8	5.8	5.7
Total	100.0	100.0	100.0

[1]*Calculated from: 'European Community,' Petroleum Press Service, XXXVIII, No. 2 (February, 1971), Table 1, 67.*

This analysis shows a marked trend of declining coal consumption and a lagging behind of the development of nuclear energy.[36] It is not surprising, therefore, to see that many believe that for decades to come oil will be the dominant feature in Western Europe's energy consumption pattern.[37] Yet, almost all of Western Europe's oil supplies are imported.

TABLE 5

Energy Forms: Percentage Share of Inland Consumption and Projection Rate, United Kingdom[1]

	1960 %	1965 %	1969 %	Forecast 1970 %	Forecast 1975 %
Coal	74.3	62.2	50.9	47.3	33.7
Oil	24.7	34.6	42.7	43.9	41.7
Natural gas	-.-	0.4	2.5	4.9	14.0
Nuclear electricity	0.4	2.0	3.3	3.9	10.6
Hydro-electricity	0.6	0.8	0.7		
Total	100.0	100.0	100.0	100.0	100.0

[1]*Calculated from: 'Britain's Energy Pattern,'* Petroleum Press Service, *XXXVII, No. 6 (June, 1970), 210.*

In 1970, net crude oil imports amounted to 586.3 million metric tons or 11.7 million bpd. This is 13.4 percent higher than the volume imported in 1969. It amounts to more than 3.5 times the level recorded in 1960.[38] Table 6 is a representative statement of the oil supplies of seven of the leading Western European countries. Together, these seven countries accounted for 91.2 percent of Western Europe oil imports in 1970.[39] An examination of the sources of oil supply of these European countries evidences that in 1970 these countries relied on the Persian Gulf area for 51.5 percent of their total oil requirements and on Libya and Algeria for another 34.6 percent, for a combined contribution of over 86 percent. The rest was provided by Nigeria, the Caribbean and the Soviet bloc. The Soviet bloc, incidentally, contributed an amount equivalent to 2.8 percent (see Table 6).

TABLE 6

Crude Oil Imports By Source – Western Europe
(Thousand Metric Tons)

Country	Year	Persian Gulf Area	North Africa	Other Africa	Caribbean Area	Soviet Bloc	Others	Total	Percentage of Change
West Germany	1967	30,081	33,848		3,714	4,064	325	72,032	
	1968	32,443	44,689		3,005	3,954		84,091	+16.7
	1969	30,158	50,039	1,950	3,883	3,493	28	89,551	+ 6.5
	1970	35,103	49,742	7,102	3,402	3,347		98,786	+10.3
France	1967	34,814	33,104		2,800	1,629		72,348	
	1968	37,381	35,878		2,363	1,554		77,176	+ 6.7
	1969	38,685	40,000	3,241	2,436	1,828	116	86,306	+11.8
	1970	44,124	44,556	7,473	2,445	1,445	120	100,163	+16.1
Italy	1967	54,468	17,259		1,981	10,615	2,116	86,439	
	1968	54,878	24,567		2,105	11,040		92,590	+ 7.6
	1969	59,214	30,653		2,218	8,890	200	101,175	+10.2
	1970	62,927	36,548	575	2,190	9,723	119	112,082	+10.8
Netherlands	1967	22,386	7,355		1,972		158	31,871	
	1968	26,321	8,225		1,235		15	35,796	+12.3
	1969	33,433	10,338	3,645	901		47	48,364	+36.5
	1970	43,903	13,317	8,457	1,274		53	67,004	+38.5
Belgium	1967	9,727	5,606		2,136		119	17,588	
	1968	14,720	5,788		2,483		403	23,394	+30.8
	1969	17,562	7,402	822	2,237	4	426	28,453	+24.0
	1970	17,500	8,000	1,000	2,500	500	500	30,000	+ 5.4

TABLE 6 (continued)

Country	Year	Persian Gulf Area	North Africa	Other Africa	Caribbean Area	Soviet Bloc	Others	Total	Percentage of Change
EEC Total	1967	151,476	97,173		12,603	16,308	2,718	280,278	
	1968	165,743	119,147		11,191	16,548	418	313,047	+11.7
	1969	179,052	138,432	9,658	11,675	14,215	817	353,849	+13.6
	1970	203,557	152,163	24,607	11,811	15,105	792	408,035	+15.3
United Kingdom	1967	44,350	15,840		9,387		5,063	74,640	
	1968	47,886	24,292		7,975		2,427	82,580	+11.2
	1969	55,957	22,353	5,220	6,390		6,650	94,570	+14.5
	1970	59,000	27,000	7,500	4,500		6,500	104,500	+10.5
Spain	1967	12,877	4,685		3,200	455		21,217	
	1968	16,648	7,935		3,398	280		28,297	+33.4
	1969	15,540	8,155	1,108	2,807	240		27,850	
	1970	17,500	9,000	1,850	2,500			30,850	+10.8
Total	1967	208,703	117,698		25,190	16,763	7,781	376,135	
	1968	230,313	151,374	15,986	22,564	16,828	2,845	423,924	+12.8
	1969	250,549	168,440		20,872	14,455	5,467	476,269	+13.2
	1970	280,057	188,103	33,957	18,811	15,105	7,292	543,385	−14.1
Percent of Total	1967	55.5	31.3		6.7	4.5	2.0	100.0	
	1968	54.3	35.7	3.4	5.3	4.0	0.7	100.0	
	1969	52.6	35.5		4.4	3.0	1.1	100.0	
	1970	51.5	34.6	6.3	3.5	2.8	1.3	100.0	

Source: 'More Crude for Europe,' Petroleum Press Service, XXXVI, No. 4 (April, 1969), Table II, 142; 'Europe's Emphasis on Crude,' Petroleum Press Service, No. 4 (April, 1970), Table II, 136; 'Big Rise in Europe's Imports,' Petroleum Press Service, XXXVII, Press Service, XXXVIII, No. 5 (May, 1971), Table II, 173.

The Demand for Oil

15

This pattern of supply structure prompted Dr Darkwart A. Rustow, a Middle East expert and professor of political science at the Graduate Center of the City University of New York, to observe that the Middle East is the only assured source from which the growing energy demand in Europe and Japan can be met in the coming decades. Rustow went on to note that the grim fact is that Western Europe and Japan have no full-fledged alternatives to dependence on Middle East oil.[40]

Japan

In 1970, Japan oil consumption hit a record average of 3,790,000 bpd realizing, therefore, a daily average increase of 540,000 barrels over the year before. While the world in general settled in 1970 for an oil growth rate of around 7.4 percent, Japan maintained a staggering upsurge in crude oil demand of 16.6 percent. Oil contributed around 70 percent of Japan's total energy requirements.[41] Japan's oil consumption was estimated to reach in 1975 a daily average of 6.2 million barrels, or up to 73 percent of Japan's total energy demands. In 1985 oil is expected to provide 68 percent. By then actual consumption would, however, reach the level of 13 million bpd[42] (see Table 7).

On the supply side, all of Japan's oil has to be imported. Table 8 indicates that out of 168.6 million kiloliters of crude oil imported in 1969, the area around the Persian Gulf provided Japan with 150.6 million kiloliters, or about 90 percent of Japan's total oil imports (see Table 8).

In summary the foregoing gives a clear picture of the oil requirements of the industrialized nations, together with their supply positions. The discussion has indicated that, at least within the foreseeable future, oil will be the dominant supplier of a surge in world energy demand. No substitute is seriously expected to take its place. Alternative sources are likely to complement oil rather than replace it.[43]

TABLE 7

Energy Forms: Percentage Share of Consumption and Projection Rate, Japan

	1968 %	1975 %	1985 %
Water power	7.8	4.5	2.5
Atomic power	0.1	2.2	9.9
Coal	23.6	18.1	16.7
Petroleum	66.5	73.0	67.8
LNG	-.-	1.0	1.4
Natural gas	1.0	0.6	1.1
Others	1.1	0.6	0.6
Total	100.0	100.0	100.0

Source: Report of the Japanese Energy Council, cited in Toshiaki Ushijima, 'Japan's Vigorous Oil Search Widens,' World Petroleum, XXXXI, No. 12 (December, 1970), Table 1, 34.

It follows, therefore, that under these assumptions the free world will consume during the 1970s and 1980s an aggregate amount of 500 billion barrels of oil – about 80 percent of the present free world reserves. If one further assumes the necessity of maintaining a production-to-reserve ratio of 1-15 to be consistent with the element of safety, the problem then becomes one of adding new oil reserves in the order of 450 billion barrels until 1990. This is equivalent to half the oil found since the industry began, or, put differently, two-and-a-half times as much as the oil produced since the industry was founded.[44]

Clearly the task is not an easy one and until these quantities are located, the Persian Gulf area with its 60 percent of the world proven oil reserves will continue to be the scene of a power struggle. For we must remember that the demand for oil is a derived demand. Oil is needed because it powers most of world industry and moves all transportation. Without it all these 'would come to a grinding halt,' according to Senator Clifford Hansen, a member of the Oil Compact Commission.[45]

17

For this reason the question of oil has always been raised with an eye to matters pertaining to the problems of national security. Considerations of national security have thus been the guiding principles in directing the search for oil. They were responsible for the emergence of a host of exceptional policies, which measures had their impact on the structuring of the oil establishment, a topic fully discussed in the following chapter.

TABLE 8

Crude Oil Imports By Source – Japan
(Thousand Kiloliters)[1]

Source	1957	Percent	1962	Percent	1968	Percent	1969	Percent
Persian Gulf Area	12,515	84.0	38,733	84.2	129,979	91.6	150,394	89.3
Saudi Arabia	6,395	43.0	10,570	23.0	28,626	20.1	29,969	17.2
Kuwait	3,120	20.9	13,443	29.0	16,791	11.9	14,326	8.5
Neutral Zone	657	4.4	5,567	12.1	17,437	12.3	18,379	10.9
Iran	613	4.1	6,425	14.0	55,384	39.1	75,169	44.6
Qatar	370	2.5	481	1.0	304	0.2	165	0.1
Iraq	1,360	9.1	2,146	4.7	1,260	0.9	146	0.1
Other	0	0.0	101	0.2	10,177	7.1	13,240	7.9
South East Asia	2,207	14.8	4,963	10.8	10,376	7.3	16,656	9.9
USSR			2,222	4.8	815	0.6	638	0.4
Others	177	1.2	88	0.2	740	0.5	898	0.5
Total	14,899	100.0	46,006	100.0	141,910	100.0	168,586	100.0

[1]*One kiloliter equals 0.863 metric ton or 6.29 barrels.*

Source: Petroleum Association of Japan, cited in Toshaki Ushijima, Japan's Vigorous Oil Search Widens,' World Petroleum, XXXXI, No. 12 (December, 1970), Table 3, 36.

Endnotes

[1] 'Why the Search Goes On,' *Petroleum Press Service*, XXXVII, No. 6 (June, 1970), 202; 'Current Analysis, Oil,' *Standard and Poor's Industry Surveys*, CXXXIX, No. 24, Sec. I (June 17, 1971), 043.

[2] 'US Oil Supply Losing Race to Demand,' *Oil and Gas Journal*, LXIX, No. 3 (January 18, 1971), 32; *Petroleum Press Service*, XXXVIII, No. 2 (February 1971), 69; 'US Dependence on Oil Imports,' *Petroleum Press Service*, XXXVIII, No. 5 (May, 1971), 170.

[3] *Petroleum Press Service*, XXXVIII, No. 2 (February 1971), 69.

[4] 'US Oil Supply Losing Race to Demand,' *Oil and Gas Journal*, LXIX, No. 3 (January 18, 1971), 32.

[5] Ibid.

[6] Ibid.

[7] Monty Hoyt, 'Oil Output on Verge of Decline,' *The Christian Science Monitor* (June 7, 1971), 1.

[8] 'North American Reserves-1969,' *Petroleum Press Service*, XXXVII, No. 5 (May, 1970), 181.

[9] Ibid., 181; 'More Domestic Oil Vital for US Health,' *Oil and Gas Journal*, LXIX, No. 7 (February 15, 1971), 50; see also, 'Basic Analysis: Oil,' *Standard and Poor's Industry Surveys*, Section 4 (April 30, 1970), 027.

[10] 'US Oil Supply Losing Race to Demand,' *Oil and Gas Journal*, LXIX, No. 3 (January 18, 1971), 32; 'How the Giants Rank,' *Oil and Gas Journal*, LXVIII, No. 30 (July 27, 1970), 173.

[11] Halford L. Hoskins, 'Needed: A Strategy for Oil,' *Foreign Affairs*, XXIX, No. 2 (January 1951), 229.

[12] 'US Energy Shortage Said Worst of Century,' *Oil and Gas Journal*, LXVIII, No. 26 (June 29, 1970), 40; *Petroleum Press Service*, XXXVII, No. 10 (September 1970), 340.

[13] 'It's Official: Foreign Oil Costs More,' *Oil and Gas Journal*, LXVIII, No. 38 (September 21, 1970), 68; Monty Hoyt, 'US Burning Its Way Toward Fuel Crisis,' *The Christian Science Monitor* (June 3, 1971), 6.

[14] Editorial, Robert W. Scott, 'Petroleum in the Year 2000,' *World Oil*, CLXXI, No. 3 (August 15, 1970), 29; 'What Project Rulison Can Mean to US Domestic Energy Supply,' *World Oil*, CLXXI, No. 6 (November 1970), 100; W.A. Bachman, 'Is the US Vastly Underestimating Its Oil Needs?' *Oil and Gas Journal*, LXIX, No. 8 (February 22, 1970), 33.

[15] Quoted in 'It's Official: Foreign Oil Costs More,' *Oil and Gas Journal*, LXVIII, No. 38 (September 21, 1970), 68.

[16] See Gene T. Finney, 'Nixon Moves to Aid Foreign Oil Talks,' *Oil and Gas Journal*, LXIX. No. 4 (January 25, 1971), 85.

[17] Quoted in W.A. Bachman, 'Is the US Vastly Underestimating Its Oil Needs?' *Oil and Gas Journal*, LXIX, No. 8 (February 22, 1971), 34.

[18] 'US Energy Shortage Said Worst of Century,' *Oil and Gas Journal*, LXVIII, No. 26 (June 29, 1970), 40; *Petroleum Press Service*, XXXVII, No. 10 (September, 1970), 340.

[19] Michel T. Halbouty, former President of the American Association of Petroleum Geologists, put the demand for oil in 1980 at 21 million bpd. 'More Domestic Oil Vital for US Health,' *Oil and Gas Journal*, XIX, No. 7 (February 15, 1971), 50.

[20] US Department of Interior, Bureau of Mines, 'The Mineral Industry of the USSR,' *Minerals Yearbook, Area Reports: International* (Washington DC: Government Printing Office, 1971), 755.

[21] Soviet Oil in the 'Seventies,' *Petroleum Press Service*, XXXVII, No. 1 (January, 1970), 3; Robert E. Hunter, *The Soviet Dilemma in the Middle East. Part II: Oil and the Persian Gulf*, Adelphi Papers, No. 60 (London: The Institute for Strategic Studies, October, 1969), 2.

[22] US Department of Interior, Bureau of Mines, op. cit., 755; 'Russian Reserves Are Inadequate,' *Petroleum Press Service*, XXXVI, No. 4 (April, 1969), 122.

[23] 'Russian Reserves Are Inadequate,' *Petroleum Press Service*, XXXVI, No. 4 (April, 1969), 122; 'Soviet Drilling Costs Up As Technology Lags Behind,' *World Petroleum*, XL, No. 7 (July, 1969), 20A; 'Growing Soviet Economic Stake in Middle East,' *Middle East Economic Digest*, XIV, No. 35 (August 28, 1970), 1010.

[24] 'Russian Reserves Are Inadequate,' *Petroleum Press Service*, XXXVI, No. 4 (April, 1969), 122; US Department of Interior, Bureau of Mines, op. cit., 755.

[25] 'Slow Decline of Soviet Exports,' *Petroleum Press Service*, XXXVII, No. 5 (May, 1970), 165; *Middle East Economic Digest*, XIV, No. 20 (May 15, 1970), 591.

[26] 'Growing Soviet Economic Stake in Middle East,' *Middle East Economic Digest*, XIV, No. 35 (August 28, 1970), 1011.

[27] 'Soviets Face Expansion Slowdown in '71,' *Oil and Gas Journal*, LXIX, No. I (January 4, 1971), 40.

[28] *Petroleum Press Service*, XXXVIII, No. 2 (February 1971), 64.

[29] Ibid.

[30] 'When Oil Flows East,' *The Economist*, CCXXXIV, No. 6594 (January 10, 1970), 51.

[31] 'Soviets Face Expansion Slowdown in '71,' *Oil and Gas Journal*, LXIX, No. 1 (January 4, 1971), 40.

[32] 'Soviet Oil Minister Sees 10 Million B/D Flow by 1975,' *Oil and Gas Journal*, LXVIII, No. 34 (August 24, 1970), 21-22.

[33] 'Western Europe: Inland Sales in 1970,' *Petroleum Press Service*, XXXVIII, No. 1 (January, 1971), 26.

[34] 'European Community,' *Petroleum Press Service*, XXXVIII, No. 2 (February, 1971), Table I, 67.

[35] 'Britain's Energy Pattern,' *Petroleum Press Service*, XXXVII, No. 6 (June, 1970), 209-210.

[36] Ibid., 211.

[37] See 'US Dependence on Oil Imports,' *Petroleum Press Service*, XXXVIII, No. 5 (May, 1971), 172; Walter J. Levy, 'Oil Power,' *Foreign Affairs*, XLIX, No. 4 (July, 1971), 664.

[38] 'Big Rise in Europe's Imports,' *Petroleum Press Service*, XXXVIII, No. 5 (May, 1971), 172.

[39] Ibid., 173.

[40] See 'US Dependence on Oil Imports,' *Petroleum Press Service*, XXXVIII, No. 5 (May, 1971), 172; 'Suez Versus Supertankers,' *Middle East Economic Digest*, XV, No. 7 (February 12, 1971), 3.

[41] 'Demand Will Rise 7.4% This Year,' *World Petroleum*, XXXXI, No. 9 (September, 1970), Table 1, 28; 'Rising Demand for Oil Met in 1970,' *Middle East Economic Digest*, XV, No. 2 (January 8, 1971), 33.

[42] 'Money Woes Crimping Japan's Ability to Meet Skying Needs,' *Oil and Gas Journal*, LXVIII, No. 31 (August 3, 1971), 57; Toshiaki Ushijima, 'Japan's Vigorous Oil Search Widens,' *World Petroleum*, XXXXI, No. 12 (December, 1970), 34.

[43] Editorial, Robert W. Scott, 'Petroleum in the Year 2000,' *World Oil*, CLXXI, No. 3 (August 15, 1970), 29.

[44] 'The Task: Find 250-450 Billion BBL,' *Oil and Gas Journal*, LXVIII, No. 26 (June 29, 1970), 35; 'Why the Search Goes On,' *Petroleum Press Service*, XXXVII, No. 6 (June, 1970), 202.

[45] Quoted in A.H. Hudkins, 'Gas and Oil,' *World Oil*, CLXXI, No. 6 (November, 1970), 15; see also Christopher Tugendhat, *Oil: The Biggest Business* (New York: G.P. Putnam's Sons, 1968), 231.

2

The Strategic Significance of Oil and Oil in the Persian Gulf

The center of gravity of the world oil industry is firmly established in the area around the Persian Gulf, which comprises the countries of Kuwait, Saudi Arabia, Bahrain, Qatar, the seven principalities of the Omani Coast (Trucial Coast) and Oman on the western shore of the Gulf, Iran on the east, and Iraq on the north. The whole area has been the scene of a vigorous and startling struggle among the powers of the world for a share in the control of its oil resources. That struggle subsided only to be resumed under different banners, using different methods.

The story of the power struggle for the control of oil will be told in the next two chapters. This section sets out the importance of oil as a commodity of strategic significance. The fact that oil in the Persian Gulf is largely controlled by the United States makes it relevant, at this level, to bring into the discussion the oil foreign policy of the United States, since that policy is instituted by reasons of national security and has its impact on the oil industry everywhere.

The backbone of modern economy is mechanization, and mechanization means oil. For oil is power. 'It is power in time of peace to develop great industrial establishments, and to transport goods and passengers on land, at sea, and in the air. In time of war oil is power to expand industry and to exert strength at great distance.'[1] This statement was made in 1951. Quite recently, Monty Hoyt of the *Christian Science Monitor* had this to say: 'Behind United States claims to be the greatest industrial nation in the world has been what seemed an endless supply of two fuel sources: oil and natural gas. Not any more.'[2] Within the context of discussing the importance of oil control, United States Director of the Office of Emergency Preparedness and Chairman of the

President's Oil Policy Committee, General George A. Lincoln, declared that 'we are a great country largely because of our supply of energy.' Lincoln went on to warn: 'Without control of that energy supply, we could become a Samson shorn of his locks.'[3]

On the strategic significance of oil, one is always reminded of Lord Curzon's remark that in the First World War 'the Allies had floated to victory on a wave of oil' and that lack of oil had been an important factor in bringing the Germans to their defeat.[4] To this end, United States Chairman of the Inter-Agency Committee on Oil during World War II, Dr Herbert Feis, says: 'Our experience was again proving how great a volume of oil was needed to fight a war, and how vital a factor it was in victory or defeat.'[5]

There are those, however, who believe that modern war technology changed much of the strategic significance of oil. They usually refer to a nuclear confrontation in which, they say, 'the oil issue becomes irrelevant.'[6] While this observation is basically correct, it nevertheless does not go far enough. It ignores the development of two related phenomena, ironically resulting from the perfection of nuclear parity among the superpowers. First, in a nuclear confrontation, nothing could really be ranked to the level of being strategic or significant, not even within the borders of the contestants. Secondly, while nuclear weapons have reasonably nullified a resort to a war, they have shifted the emphasis from one of winning over a rival to one of disciplining an associate, an ally, or a member of a bloc. And no instrument is so powerful to bring pressure to bear on an associate as the one which could bring its economy to a grinding halt – the control of its energy supply.

We know that Western Europe and Japan are heavily dependent on oil from the Persian Gulf. If, therefore, the control of their oil supply sources falls into the 'wrong hands,' it could be 'a sword of Damocles hanging over the free world.'[7] Within the context of discussing the Soviet drive in the direction of the Persian Gulf, Ruth Sheldon Knowles, petroleum specialist, remarked that 'the Soviets can envision their influence extending to the point of having an economic stranglehold on Western Europe and Japan.'[8] Similarly, the Soviets as discussed later, have an interest in controlling the countries of Eastern Europe, insisting on providing these countries with the bulk of their oil needs.[9]

It appears, therefore, that the management of international relations today has its own instruments which have been adapted to the requirements of the nuclear age. Professor T.B. Millar, the Director of the Australian Institute of International Affairs, warns of Soviet control over the Persian Gulf and states that 'the diplomatic, strategic and economic leverage which could come from increased Soviet influence over the principal oil producers is not something which the West can lightly concede.'[10] Bemoaning the decline of the oil reserves in the United States, Dr Charles A. Heller, petroleum specialist, wished to see the Prudhoe Bay oil fields as huge as those in the Middle East so that 'the hegemony of the Middle East with all its oil reserves of over 268 billion barrels could come to an end.' If that were to happen, says Heller, 'more will be changed than the economics of oil.' Heller then went on listing the changes: 'The dominance of the North American continent will be strengthened and the international power balance will be vitally affected. For such is the impact of oil on the relation of nations, today and tomorrow.'[11]

The strategic significance of oil has been behind the United States' world search for oil, and oil today is the only product listed under the national-security clause. In 1944 as in 1919, the United States government was seriously disturbed over the exhaustion of its domestic oil reserves as a result of her engagement in two wars. 'Vigorous and startling measures were taken for the purpose of assuring full access to and development of foreign supplies – especially those in the Middle East.'[12]

Accordingly, and largely as a result of these American efforts, which were preceded by a British initiative, the largest petroleum stores in the world were discovered in the Persian Gulf area. By the end of 1970, published proven oil reserves were put at 47,823 million tons (see Table 9). This is over 269 billion barrels and represents around 60 percent of world oil reserves. By comparison, oil reserves in the United States, excluding the Prudhoe North Slope, were put at eight percent of the world's proven reserves.[13] If the oil reserves of the United States and Soviet Union were to be excluded for their negligible contribution to the international movements of oil, the oil reserves in the Gulf would reach the level of 75 percent of the rest of the world.[14]

TABLE 9
'Published Proven' Oil Reserves in the Gulf Area (Million Tons)

	End 1963	End 1967	End 1969	End 1970	1970 percent of world total
Iran	5,007	6,000	7,535	9,590	11.5
Iraq	3,421	3,219	3,767	4,384	5.2
Kuwait	8,773	9,589	9,316	9,193	11.0
Neutral Zone	1,460	1,849	1,781	3,521	4.2
Qatar	384	514	754	589	0.7
Saudi Arabia	8,118	10,233	11,500	17,604	21.0
Others	1,348	2,926	4,04	22,942	3.5
Total	28,581	34,330	38,695	47,823	57.1

Source: The Middle East and North Africa, *1970-71 (17th edn., London: Europa Publications Limited. 1970), 44;* The Middle East and North Africa, *1971-72 (18th edn., London: Europa Publications Limited, 1971), 38.*

A special feature of the Gulf is the routine way in which oil reserves are piled up. While the industry was jubilant over the addition of five to ten billion barrels of oil reserves wrested from the fierce jaws of Alaska, John Jamieson, Chief Executive of Standard Oil of New Jersey, made the observation that the Arabian American Oil Company (Aramco) had 'as a routine matter' added 8.5 billion barrels to its Saudi Arabian reserves in 1968 alone.[15]

More importantly, especially for the power struggle, is the Gulf area's undiscovered potentialities. As yet, the area has barely been scratched, and the off-shore search is in its early stage of development. The area is now the scene of a renewed struggle among the powers of the world for a share in its resources.[16]

Oil production in the Gulf amounted in 1970 to nearly 680 million tons. This is nearly 30 percent of the world total production (see Table 10). The region, therefore, retained its first place position, which it secured in 1966 as the largest oil-producing area in the world. Production in 1969 amounted to 611 million tons and was 18 percent higher than the 519 million tons produced by the United States, the world's second largest oil-producing country. The Soviet Union, with its production of 358 million tons, came third.[17]

Ninety percent of 1968 Gulf production was contributed by four countries: Iran, Iraq, Kuwait, and Saudi Arabia.[18] In 1968 the Gulf region was responsible for over half the oil that entered international trade. North Africa's contribution was put at 13 percent and Venezuela's at 20 percent. The rest was provided by Nigeria, Indonesia, and Canada.

The abundance of resources in the Gulf is matched by the cheapness of the cost for exploration, development, and production. Oilfields in the Gulf region are often huge, and the oil is held under considerable pressure so that few wells need be sunk to tap a large area. All the wells are natural flowing and require no artificial pumping. Average production per well in 1970 was put at 10,000 bpd.[19]

By way of comparison, at the end of 1967 there were about 570,000 wells in the United States with an average production of 15.2 bpd. Of the total wells available in 1969, 358,650 were the so-called stripper wells with an average production range from 0.29 bpd in the state of Pennsylvania to a high of 7.97 bpd in the state of Colorado, giving, thereby, a national average of 3.47 bpd.[20]

The difference between these geological structures showed itself in the amount of costs incurred, both in terms of current operating costs and in terms of capital expenditures. In Kuwait, for example, the cost of producing one barrel averaged $0.062 in 1967. In the United States it was put at $1.31. Other estimates put the cost

29

of finding, developing and producing a barrel of crude in the Persian Gulf and Libya at between 10 and 20 cents. The corresponding cost for the United States was put at $1.40. A third estimate put the average production cost of a barrel in the Persian Gulf at $0.12 while that of the United States was $2.50.[21]

TABLE 10

Oil Production in the Gulf Area[1]
(Thousand Tons)

	1966	1967	1968	1969	1970	1969-70 Percent Change
Iran	105,855	129,298	141,791	168,235	190,000	+12.9
Saudi Arabia	119,456	129,300	140,998	148,839	175,500	+17.9
Kuwait	114,355	115,169	122,085	126,549	138,000	+6.5
Iraq	68,011	59,981	73,848	74,700	75,000	+1.2
Abu Dhabi	17,313	18,125	24,006	28,761	32,800	+14.9
Neutral Zone	22,341	22,618	22,827	23,502	27,000	+14.9
Qatar	13,845	15,479	16,363	17,341	17,000	-1.0
Oman		2,800	12,068	16,069	16,400	+2.0
Dubal				523	4,200	
Bahrain	3144	3,443	3,768	3,795	3,800	
Gulf total	464,320	496,213	557,754	611,314	679,700	+11.0
World total	1,641,185	1,757,555	1,975,315	2,134,531	2,334,000	+9.4
Gulf percent of world total	28.3	28.2	28.2	28.6	29.1	

[1]*Calculated from* Middle East Economic Digest, *XV, No. 2 (January 8, 1971), 33.*

These fantastic cost differences prompted oilmen to admit that 'neither Alaska's big Prudhoe Bay fields, nor others in Canada or the North Sea can be competitive in the foreseeable future with the low costs and huge reserves of the Middle East.'[22] For in the Middle East, 'A well can "pay out" in days and sometimes hours. For example, one well in Gachsaran field, Iran, flows at a sustained rate of 100,000 barrels of oil every day.'[23]

This position of the Gulf area was known to American officials long before an American mission under the leadership of Everette de Golyer was sent to the scene in 1944 to report on the immensity of the great oil reserves of that part of the world.[24] In 1970, the United States controlled over 60 percent of the oil reserves of the Gulf area, nearly in the proportion shown in Table 11. The British came second in the magnitude of their holdings, followed by the British-Dutch, and then the French.

Despite the magnitude of its holdings, the United States embarked upon a policy that appears almost without justification or, at least, out of proportion. The essence of this policy is that the United States should, in the name of national security, hold on to controlled reserves overseas, but supply most of the United States oil needs from the United States itself. This despite the fact that United States oil is almost twenty times as costly to produce per barrel as that of the Persian Gulf.

In practice, however, this policy boiled down to one of maintaining self-sufficiency and benefiting from the marketing of foreign oil to Europe, Japan, and the rest of the world. The policy is seen, therefore, as a move in the direction of perpetuating the uneven distribution of wealth between nations. It is reminiscent of Gunnar Myrdal's vicious circle in the relation between rich lands and poor lands. The policy had been the cause of unnecessary friction, yet it provided very little in the way of enhancement to the national security of the United States.

TABLE 11

Major Oil Reserve Holdings in the Gulf Area

	United States								Britain	Anglo-Dutch	France
					Percent						
	Standard N.J.	Standard Calif.	Texas Company	Socony Mobil	Gulf Oil	Getty Oil	American Independent[1]	Iricon Agency[2]	British Petroleum	Royal Dutch Shell Group	Campagnie Francaise
Saudi Arabia	30	30	30	10							
Kuwait					50				50		
Iran	7	7	7	7	7			5	40	14	6
Iraq	11.875			11.875					23.75	23.75	23.75
Neutral Zone						50	50				
Qatar	11.875			11.875					23.75	23.75	23.75

[1] American Independent Oil Co. is owned by Phillips Petroleum 37%; Signal Oil and Gas 33.6%; Ashland Oil 14.13%; Sunray DX 2.95%; and other investors some 12.3%.

[2] Iricon Agency is owned by Atlantic Richfield 1.67%; American Independent Oil Company, Getty Oil and Signal Oil and Gas, 0.833% each; and Standard Oil of Ohio and Continental Oil, 0.471% each.

Source: 'Basic Analysis Oil,' Standard and Poor's Industry Surveys, Section 2 (December 11, 1969), 0 55.

Early in the 1950s, cheap oil from foreign sources found its way to United States markets. Voices were raised as to the danger of creating dependence on foreign oil and putting economic pressure on local producers which would cause them to abandon their oil reserves and exploratory efforts. In July 1957, a voluntary curb was put on the importation of foreign oil. Two years later, the United States turned into a net oil importer and on March 11, 1959, the Eisenhower Administration instituted the mandatory curb of restricting the importation of foreign oil to a quota of 12.2 percent of domestic production.[25]

In the process, oil production in the Persian Gulf area was kept at a level disproportionate to the area's position in world reserves for reasons of America's national security. Production was held to a minimum – just enough to supplement the United States domestic supply and prevent foreign countries from canceling the concession agreements. To substantiate this, the United States Navy oil expert during the Second World War, Roger R. Sharp, said:

> Imports of foreign petroleum relate to national security and preparedness in several respects. By supplementing the domestic supply in peacetime, domestic reserves are conserved for an emergency. We also need to maintain imports at a level which will protect United States oil companies' foreign concession agreements. For unless a *minimum* production in such foreign countries is preserved, there is a danger that concession agreements may be cancelled.[26] (Emphasis added.)

In 1966, the Gulf area provided the United States with 314,900 bpd, or 12.8 percent of the US petroleum import market. Imports then started to decline each year until, in 1969, they reached a level of 182,000 bpd, or 5.9 percent of United States total imports, a fall of 41.9 percent from the 1966 level (see Table 12). The biggest decline took place immediately after the Arab–Israeli war of 1967. Thereafter, the area started to recover its position until in

1969, when a new low was registered. Iraq at that time was dropped entirely as a United States importer while Saudi Arabia, Iran and Abu Dhabi hit new lows. Kuwait and the Neutral Zone appeared to be slight gainers.[27]

TABLE 12
United States Petroleum Imports by Source (Thousand Barrels Per Day)

	1966	% of Total	1969	% of Total	1966-69 Change
Latin America	1,588.9	64.3	1,681.2	54.5	+5.2
Canada	387.3	15.7	612.2	19.9	+58.1
Persian Gulf	314.9	12.8	182.6	5.9	-41.9
Africa	88.5	3.6	222.8	7.2	+150.6
Far East	57.0	2.3	89.6	2.9	+57.2
Europe	26.4	1.1	171.5	5.6	-.-
Virgin Islands	5.5	0.2	122.0	4.0	-.-
Total	2,468.5	100.0	3,081.9	100.0	

Source: Oil and Gas Journal, *LXVIII, No. 29 (July 20, 1970), 25.*

Behind this policy there exists a suspiciousness on the part of the United States as to the dependability of these oil-producing countries to perform as a source of energy supply.[28] Thus, while these countries trusted the United States and let her control the larger portion of what is almost their only generator of government revenues, the United States, in turn, penalized them by curbing their oil exports on the basis that they do not deserve the trust. Of course, one cannot fail to see the strained Arab–American relations that came out as a result of America's support for Israel. The issue, therefore, should be put differently. It is not the Arabs whom the

United States cannot trust; it is the United States who should adjust her policy to reflect doing what is in her national interest.

Nevertheless, the antagonism has been going on for so long with the quota system in fact adding very little to buttress the United States national security, at the same time costing every American family of four between $102.32 and $258 per year.[29]

For these reasons a Cabinet Task Force was set up in March 1969, by President Nixon to review the viability of the import control. The guiding principles of the Task Force had been, firstly, the protection of the United States national security against possible foreign oil supply interruption; and, secondly, the prevention of damages to the domestic oil industry as a result of the relaxation of import control.[30]

Before reaching any conclusion on the effect of import relaxation, the Task Force decided to first assess the cost of the existing quota program. It was put at '$3.0-$3.5 billion per year of transfer payments from one sector of the economy to another plus efficiency losses in production and transportation of about $1.5-$2.0 billion per year, including losses caused by "market-demand prorationing" as practiced by the principal producing states.'[31]

The Task Force made the observation that the risks to the United States national security for which these costs were incurred do not in the main concern the functioning of United States armed forces. A hostile superpower might be tempted to interfere with US oil tanker movements in the high seas. But, said the Task Force, such an action could not go on indefinitely; it would be either 'settled or escalated.'[32]

With this in mind, the Task Force assessed the kinds of risks to the United States national security that could result from the eruption of war between the Arabs and the Israelis and the risks that could arise from a concerted oil boycott by the oil-producing Arab countries.

With regard to the first risks, the Task Force observed that with the coming of the supertankers and the avoidance of the Suez Canal – a war zone – there is little chance for an Arab–Israeli war to cause interruption of the flow of oil from the Persian Gulf area.

Besides, it was noted that in the past there had been three wars between the Arabs and the Israelis during which Europe – an area totally dependent area on Middle Eastern oil – suffered only minor inconveniences.

There remains the risk to the United States security from a concerted action by the oil-producing Arab states. Here, at least three factors could be listed as working against the Arab ability to resort to such an action. The first and most important is the one raised by the Task Force, i.e., the complete dependence of the governments of the oil-producing countries on revenue from their oil. On this point the Task Force stated that 'there are factors of self-interest at work in this region to limit the duration of any such supply interruption.'[33] Second, for the Arabs to have any effective boycotting against the United States they must also have the compliance of Iran, Nigeria and, possibly, Venezuela – a contingency which they could not marshal in their embargo of 1967. Finally, the Arabs have been known for their fragmentations. Their weakness resides precisely in their inability to take a unified stand on almost any issue, let alone an issue the magnitude of an oil embargo. An oil supply interruption by only one or two Arab states is almost too insignificant to cause the United States worry over her national security.

We now consider the second principle that guided the Task Force, namely, ensuing damages to the domestic industry from oil import relaxation. It must be recalled that in the background of this principle is the preservation of certain production-reserve ratios said to be concomitant with the element of safety. The supporters of this principle maintain that without the mandatory restriction, cheap foreign oil would find its way to United States markets, thus causing the abandonment of exhausted fields and exploratory – both tantamount to oil reserve accumulations.

The quota system, and before it the voluntary import restrictions, had been in existence for over twelve years, and yet the United States domestic oil reserves have been dwindling over those years. Import quotas, as such, were challenged by the Antitrust Division of the Department of Justice on the grounds that these quotas 'do nothing to preserve the nation's domestic oil reserves.' The Antitrust Division went on to say:

> Reserve productive capacity is maintained, if at all, by state regulatory actions aimed primarily at other objectives, such as conservation. The resulting hodge-podge of Federal and State regulations seems ill-adapted for achievement of a coherent program designed to provide the country with sufficient emergency oil reserves.[34]

As a result, General Lincoln, a member of the President's Cabinet Task Force on Oil Import Control and Chairman of the newly established President's Oil Policy Committee, testified before the House of Representatives Subcommittee on Mines and Mining, 'there is no question ... but that oil imports are going to have to increase. ... I think there is unanimous agreement on that.'[35]

Furthermore, the quota system has been criticized for what it does in insulating the domestic market from the competitive pressure of world price and the tendency of the quota to breed inefficiency with the resultant loss of competitive advantage – a factor in the development of United States balance of payment difficulties.

For all these reasons, a 400-page report was prepared by the Cabinet Task Force, suggesting, in a nutshell, a switch from the quota system to tariff arrangements. In its recommendations, the Task Force did not forget that the Middle East is a volatile area and that caution is therefore in order. The Task Force therefore injected a preferential measure to its recommendations. It maintained that although there should be a change from quotas to tariffs, there should also be a 'phased-in adoption of preferential tariffs with an Eastern Hemisphere security adjustment.' The scheme called for the exemption of Canadian and Mexican oil from the entire tariff program and the giving of preferential tariff treatment to oil from the Western Hemisphere in general so as 'to neutralize the advantages of Eastern Hemisphere oil without exerting undue competitive pressure on US production.'[36]

Despite all the reasoning and the precautionary arrangements provided by the Task Force for the substitution of tariffs for quotas, the measure failed to pass the US Congress. According to Irwin Knoll, passage threatened a substantial reduction in the oil industry's profits.[37]

The level of oil production in the United States is established by regulatory agencies in the major producing states in volumes known in the trade as 'allowable.' This allowable is usually 'nominated' by the major oil companies. The major companies, in other words, 'tell the [regulatory] agencies how much crude they require in any given month.' The oil companies then turn around to demand and get a tight import control specified by quotas. They, therefore, put a ceiling on the supply side of the transaction and, accordingly, keep the prices at their high level.[38]

Now, if tariffs were to be substituted for quotas, independent (small) oil companies would, in the interest of capturing a segment of the United States oil consumption market, be lured to flood the market with oil from their cheap foreign resources. These independents will achieve their objectives by engaging in competitive price cuttings. Of course, the independents' access to voluminous quantities and marketing facilities is decidedly modest in comparison to the resources under the major companies' command. Yet, it is always the amount in the margin that has its impact on prices. With the elasticity of demand near one, reduction in prices would not lead to a proportionate increase in the quantity consumed. It would, instead, lead to a reduction in the profits now realized.[39] This the major companies do not want.

Hinging their case on problems of national security, the oil companies, therefore, pooled their efforts and for many weeks 'bombarded Congress and the White House with demands that the quota system be retained.' Consequently, the House Ways and Means Committee overrode the Task Force recommendations and required the President to continue to use quotas in limiting oil imports. It was also recommended that imports from the Eastern

Hemisphere should be limited to five percent of the United States import market.[40]

Thus, the oil lobby won its case. Senator William Proxmire, an ardent critic of the oil industry, put the industry fiscal advantage at over 9.3 billion a year from subsidies and tax preferences in the forms of depletion allowance, foreign tax credits for disguised royalty payments, expenses for intangibles and higher prices because of the quota. Senator Proxmire described the oil lobby as 'the most potent and most richly rewarded lobby in Washington.'[41]

In an article heavily critical of the oil industry, *Business Week* stated that the oil industry 'has come to be dominated by a handful of giants that are integrated in all phases of the oil business.' According to Dr Malcolm Caldwell, a British scholar and a lecturer at the School of Oriental and African Studies in London, this kind of business is 'extremely dangerous, because these vast industrial empires answer only to a handful of the leading shareholders.'[42] Americans have no justification to fear the bigness of the oil companies were it not for the fact that these 'companies have always had a knack for running things their own way.'[43]

What worries the anti-trusters further is the direction in which these companies are leading themselves – to become energy establishments. They are in the natural gas business, and this is an area which is usually exploited along with oil. They have been buying heavily into the coal industry. For example, Humble Oil Company, a subsidiary of Standard Oil of New Jersey, 'bought up hundreds of thousands of areas of coal land in the Midwest and Western plains states.' The implication of this acquisition had been touched upon by Walker B. Comeggs, Chief Deputy of the Anti-trust Division. The acquisition has been noted for the ensuing lack of competition among different energy producers and the effect of this development on prices. It is believed that it was not a coincidence for the price of coal to go up from 28 cents for a million BTUs (British thermal units) to a range of between 48 and 75 cents in a period of 18 months after the acquisition took place.[44]

To top it all, the oil companies moved to acquire still another source of energy: uranium reserves. Their holdings of this source of energy are put at 45 percent of the known reserves in the United States. Other estimates put the amount of uranium reserves controlled by the oil companies at 80 percent.[45]

The tendency, therefore, is for these companies to control the source of energy, whether it is oil, coal, or the atom. 'If,' says the *Courier Journal*, 'the oil industry can get control over the fuel from which electricity is generated, it will have the country by the throat.' For this reason Senator Proxmire describes his fight with the oil companies as a fight against an octopus. Dr Caldwell describes them as 'great powers' in today's world. He ascribes to them the management of foreign policy.[46]

It is through the medium of these international oil companies that the struggle for world petroleum dominance is carried on. Each group of these companies is actively supported by its respective government: 'For the Anglo-Dutch companies, the Foreign Office of the British Government; and for the United States companies, the State Department.'[47] This support is to be discussed next.

Endnotes

[1] Halford L. Hoskins, 'Needed: A Strategy for Oil,' *Foreign Affairs*, XXIX, No. 2 (January, 1951), 229.

2 Monty Hoyt, 'Oil Output on Verge of Decline,' *Christian Science Monitor* (June 7, 1971), 1.

[3] Quoted in 'Lincoln Pleads for US Energy Plan,' *Oil and Gas Journal*, LXIX, No. 4 (January 25, 1971), 90.

[4] Quoted in Christopher Tugendhat, *Oil: The Biggest Business* (New York: G.P. Putnam's Sons, 1968), 113; Joseph E. Pogue, 'Must An Oil War Follow This War?' *Atlantic Monthly*, CLXXIII, No. 3 (March, 1944), 41.

[5] Herbert Feis, 'Oil For Peace Or War,' *Foreign Affairs*, XXXII, No. 3 (April, 1954), 417.

[6] O.M. Smolansky, 'Moscow and the Persian Gulf: An Analysis of Soviet Ambitions and Potentials,' *The Princeton University Conference and Twentieth Annual Near East Conference on Middle East Focus: The Persian Gulf, October 24-25, 1968*, T. Cuyler Young (ed.), Princeton University Conference, 1968, 153-4.

[7] Elston R. Law, 'Comment,' *The Princeton University Conference and Twentieth Annual Near East Conference on Middle East Focus: The Persian Gulf, October 24-25, 1968*, T. Cuyler Young (ed.), Princeton University Conference, 1968, 189.

[8] Ruth Sheldon Knowles, 'A New Soviet Thrust,' *Mid East: A Middle East and North Africa Review*, IX (December, 1969), 9.

[9] 'When Oil Flows East,' *The Economist*, CCXXXIV, No. 6594 (January 10, 1970), 51.

[10] T.B. Millar, 'Soviet Policies South and East of Suez,' *Foreign Affairs*, XLIX, No. 1 (October, 1970), 79; 'Russia Drives East of Suez,' *Newsweek* (January 18, 1971), 27.

[11] Charles A. Heller, 'American Petroleum Fortress?' *World Petroleum*, XL, No. 6 (June, 1969), 73.

[12] Feis, op. cit., 416; see also Pogue, op. cit., 41.

[13] 'Basic Analysis, Oil,' *Standard and Poor's Industry Surveys*, Section 4 (April 30, 1970), O 26-O 27.

[14] Tugendhat, op. cit., 165.

[15] Quoted in John K. Cooley, '10-Power Oil Talks Set Precedent,' *Christian Science Monitor* (January 29, 1971), 9.

[16] *The Middle East and North Africa, 1969-70*, (16th edn., London: Europa Publications Limited, 1969), 39; 'Gulf Oil Exploration Speeds Up,' *Middle East Economic Digest*, XIV, No. 29 (July 17, 1970), 857.

[17] *The Middle East and North Africa, 1970-71*, (17th edn., London: Europa Publications Limited. 1970), 39 and 44.

[18] William F. Todd, 'The Impact of Oil on Middle East Economies,' *World Petroleum*, XL, No. 1 (January, 1969), 38.

[19] Brian C. Hague, 'Sabiriyah Raises Kuwait Production,' *World Petroleum*, XLI, No. 11 (November, 1970), 58.

[20] Heller, op. cit., 41; 'Stripper-well Survey Shows Production Edging Down Slightly,' *Oil and Gas Journal*, LXIX, No. 4 (January 25, 1971), 132.

[21] Tugendhat, op. cit., 245; Thomas O'Hanlon, 'Mobil Oil "Squarely in the Middle East,"' *Fortune*, LXXVI, No. 3 (September, 1967), 88; Heller, op. cit., 41.

[22] Quoted in John K. Cooley, '10-Power Oil Talks Set Precedent,' *Christian Science Monitor* (January 29, 1971), 9.

[23] Robert W. Scott, editorial, 'Petroleum in the Year 2000,' *World Oil*, CLXXI, No. 3 (August 15, 1970), 34.

[24] US, Congress, Senate and House, Select Committee on Small Business of the Senate and the House of Representatives, *A Report: The Third Petroleum Congress*, 88th Congress, 2nd Sess. (Washington DC: Government Printing Office, 1952), 16.

[25] Roger R. Sharp, 'America's Stake in World Petroleum,' *Harvard Business Review*, XXVIII, No. 5 (September, 1950), 26; US, Congress, Senate, Subcommittee on Antitrust and Monopoly of the Committee on Judiciary, *The Petroleum Industry, the Cabinet Task Force on Oil Import Control*: Majority and Minority Recommendations, Hearings, Part 4, 91st Cong., 2nd Sess., March 3 and 26, 1970 (Washington DC: Government Printing Office, 1970), 1780. (Henceforth: *The Cabinet Task Force.*)

[26] Sharp, op. cit., 38.

[27] Dibrell DuVal, 'Important Shifts Occur on US Imports Scene,' *Oil and Gas Journal*, LXVII, No. 29 (July 20, 1970), 26.

[28] See M.A. Adelman, 'Security of Eastern Hemisphere Fuel Supply,' working Paper, Department of Economics, No. 6 (Cambridge: Massachusetts Institute of Technology, December, 1967), 1-9.

[29] Erwin Knoll, 'The Oil Lobby Is Not Depleted,' US, Congress, Senate, Subcommittee on Antitrust and Monopoly of the

Committee on Judiciary, *The Petroleum Industry, The Cabinet Task Force on Oil Import Control: Majority and Minority Recommendations, Hearings*, Part 4, 91st Cong., 2nd Sess. , March 3 and 26, 1970 (Washington DC: Government Printing Office, 1970), 1885.

[30] *The Cabinet Task Force*, 1780.

[31] Ibid.

[32] Ibid., 1781.

[33] Ibid.

[34] Quoted in Knoll, op. cit., 1886.

[35] US, Congress, House, Subcommittee on Mines and Mining of the Committee on Interior and Insular Affairs, *Oil Import Controls, Hearings*, Serial No. 91-17, 91st Cong., 2nd Sess., March 9, 10, 16 and 17, April 6, 7, 23 and 24, 1970 (Washington DC: Government Printing Office, 1970), 18.

[36] *The Cabinet Task Force*, 1782.

[37] Knoll, op. cit., 1886.

[38] 'Oil Is Taking Over the Energy Business,' *Business Week*, No. 2149 (November 7, 1970), 55.

[39] See Gilbert Burck, 'World Oil: The Game Gets Rough,' *Fortune*, LVII, No. 5 (May, 1958), 125-128 and 186, 188, and 193, for similar behavior of the prices when the independents first penetrated Europe late in the 1950s.

[40] Knoll, op. cit., 1887; 'Oil Import Quotas Win Votes,' *Oil and Gas Journal*, LXVIII, No. 31 (August 3, 1970), 56; 'Lid Urged for Shaky Oil Sources Abroad,' *Oil and Gas Journal*, LXVIII, No. 32 (August 10, 1970), 80.

[41] Quoted in Richard L. Strout, 'What Oil Quotas Cost US Families,' *Christian Science Monitor* (February 2, 1972), 3.

[42] 'Oil Power,' *Christian Science Monitor* (February 24, 1971), 11.

[43] *Business Week*, op. cit., 54; Robert Engler advanced the thesis that the oil industry constitutes a private government with an overwhelming power to influence political decisions. See his book, *The Politics of Oil: A Study of Private Power and Democratic Direction* (New York: The Macmillan Company, 1961).

[44] Philip W. McKinsey, 'US Trustbusters Scan Oil-Industry Operations,' *Christian Science Monitor* (January 6, 1971), 12; 'Opportunity for Oil,' *Christian Science Monitor* (February 24, 1971), 11.

[45] McKinsey, op. cit., 12; *Business Week*, op. cit., 54.

[46] Quoted in 'Who Controls Electric Energy Controls Us,' *Christian Science Monitor* (August 7, 1971), 18; *The Cabinet Task Force*, 1882; 'Oil Power,' *Christian Science Monitor* (February 24, 1971) 11.

[47] Sharp, op. cit., 27.

3

The Story of the Power Struggle
for the Control of Oil

A Prelude

The unfolding of the twentieth century witnessed the first exploitation of oil discovery in the Persian Gulf area. The British at that time had absolute hegemony over the area and were more determined than before to prevent the French, the Russians, the Germans, and even the Americans from getting a foothold in the Gulf.

The British position in the Gulf is a development of the past which has little connection to the presence of oil. Britain regarded the Persian Gulf as one of the main bastions on the route to India. Its occupation was sought for securing British lines of communication between Europe and the East.

Britain made her first entry into the Gulf area in 1622 when she captured the Island of Hormuz from the Portuguese. Portugal was the first European power in modern history to occupy posts in the Gulf. She captured the capital of Oman in 1508 and built a factory in Muscat and maintained a naval base 'for controlling trade in the Persian Gulf area' for almost 150 years. The Portuguese were finally driven out by the 'local Arabs' in 1650.[1] (Incidentally, this act was here referred to as having been committed by *local Arabs*.)

The defeat of the Portuguese put the British face to face with the Dutch. There developed a severe rivalry between these two powers for supremacy on the Gulf shores. The British, however, were able to assert their ascendancy by bringing the Dutch presence to its end in 1766.

By the close of the eighteenth century, Britain found herself again facing the French advancement toward the East. Napoleon's invasion of Egypt and the advance of the French forces in the direction of Palestine hurried the British to conclude treaties with the Sultan of Muscat (Oman) and the Shah of Persia (Iran). The treaty with the Sultan provided that the Sultan should not 'give any aid to the French.' The treaty with the Shah provided that 'the Shah should not receive French agents and would do his utmost to prevent French forces from entering Persia.'[2]

The opening of the nineteenth century witnessed Britain's complete elimination of her European rivals. Accordingly, the British continued affirming their influence and building their commercial enterprises over the Gulf water. While British forces were busy liquidating the European presence, East India Company, a British concern, was busy establishing factories, depots, and mail route terminals almost everywhere on the Gulf shores – in Bandar Abbas, in Basra, in Busire, in the town of Kuwait and, incidentally, in Muscat, where the Portuguese first had their factory.[3]

Under the direction of the East India Company, the water of the Gulf was mapped and buoyed and mail routes through the Euphrates were established. For two-and-a-half centuries this British private organization was accorded military protection and was 'the dominating commercial, military and political factor in the Gulf and its seaward approaches.'[4]

The East India Company took direct charge of the Gulf until 1858. Thereafter, control passed to the British government in Bombay until in 1873 it was vested in the colonial authority of India. When India got its independence in 1947, the control of the Gulf passed to the Foreign Office in London.[5]

This British military and commercial domination, however, was challenged by the Qawasim Arabs of the Omani Coast (called by the British the Pirate Coast until 1853 when, after the conclusion of the maritime truce, it was renamed the Trucial Coast). These

Arab tribes were wedded together by a new movement of Islamic revival.[6]

By its enemies, this Islamic movement is considered fanaticism. It was seen as an urge to pillage and plunder. The movement was further denigrated by naming it after its reformer (not founder) Sheikh Mohammad Ibn Abdalwahhab. It is therefore called the 'Wahhabi' movement.[7] The fact is that this movement is a direct campaign to a return to the purity and clarity of Islam. 'The Wahhabi creed,' says Richard Coke, 'is a sincere and not ineffectual effort at reformation, but it does not go far enough; it is right in going back to the prophet, but wrong in insisting on the letter instead of the spirit of the law.' In his *Area Handbook for Saudi Arabia*, Norman C. Walpole attests to the fact that Islam, and more particularly its Wahhabi form, is the most important unifying force in Saudi Arabia.[8]

It was this unifying force that the British feared most and were determined to crush. The religious movement, launched in the middle of the eighteenth century, had, by 1800, Kerbela and Najaf in Iraq, Damascus in Syria, and Mecca and Medina in Hejaz, following a central authority in the center of Arabia. Eastward, the movement spread to the Al-Ahsa province on the Gulf, the Buraimi oasis, and the Coast of Oman. In just a few years the people of that area 'adopted the Wahhabi practice of Islam.'[9]

Now an Arab community, united together by this movement toward Islamic revival and having under its command a fleet composed of 'sixty ocean dhows and eight hundred small craft and manned by almost twenty thousand men,' was carrying its commercial activities to the east to the coasts of India, to the north to the coasts of Persia and Iraq, and to the west to the coasts of Africa and the Island of Zanzibar.[10] These commercial activities were impinged upon by a British company which was accorded government protection. The Arabs decided to resist this intrusion in the same way they had resisted the Portuguese before. This time, however, the resistance was called piracy.

In their first engagement with the East India Company's warships, the Qawasims attacked the twenty-four-gun *Mornington* and the six-gun *Fury*. In 1808 the Qawasims captured the *Minerva*, a British merchant ship, and incidentally delivered unharmed the only female passenger on that ship. 'The Qawasims,' says Sanger, 'became Wahhabis in 1800; ... every few months after that saw some British vessel captured or forced to flee for its life.'[11]

The British, therefore, decided to act. In 1809, they sent a large fleet to destroy and burn the Arab village of Ras al-Khaima on the Omani Coast. The expedition, however, was only a setback to the Arabs, for they retreated inland and came back 'with even greater ferocity.'[12]

Ten years later, in 1819, the British saw a golden opportunity when they learned the central authority in Istanbul regarded this Islamic movement in Arabia as an attempt for secession. The authority in Istanbul, therefore, incited the Khedive of Egypt to obstruct the movement. An Egyptian expedition was mobilized and it was able to overrun the center of the Islamic movement in the core of Arabia. The Qawasim Arabs on the coast therefore lost their support from that center. The British, meanwhile, realized that without forces on the ground their campaign could end in the same way the 1809 expedition ended – with a setback. Not wanting to involve their own soldiers in this kind of ground war, they persuaded the Sultan of Muscat to provide them with four thousand troops. Thus, 'while the British ships fired from the sea, the four thousand troops of the Sultan attacked the pirates from the rear. After six days of hard fighting the rebuilt stronghold of Ras al-Khaima fell and was soon followed by the other towns of the Pirate Coast.'[13]

Although there had been sporadic attempts towards the unifying of the Arabian Peninsula during the course of the nineteenth century, the real attempt had to wait for King Abdalaziz to return from Kuwait in 1901, to capture Riyadh in 1902, and to begin the unification of the country in what is now known as the Kingdom of Saudi Arabia. Even then the British were on full alert not to let the course of history take its rightful turn. Had the British let the movement accomplish its goal, there would not have been city-states sprinkling the shores of the Gulf. For, according to Sir

Rupert Hay, Political Resident, Persian Gulf, in 1941 and 1946, 'with a few exceptions all the Gulf Rulers had a deep reverence for the late King Abdalaziz Ibn Saud and regarded him more or less as their father. They undoubtedly respect his successor.'[14] Sir Rupert went on to say:

> This does not mean that they will not put up a stiff resistance
> to any attempted Saudi encroachment, but apart from this
> they will do everything possible to avoid giving offence.[15]

This statement was not put to test, however. For when unity under the leadership of King Abdalaziz was put into effect, it proved beneficial for every part of the Kingdom. The fact is that, 'The emergence of a unified political entity was dangerous for British interests on the Arab coast.'[16] Any further attempts at unification were, therefore, suppressed.

The suppression of the Qawasims, or the pirates as the British liked to call them, was followed by the conclusion of a series of treaties aimed at giving the British a free hand in the Gulf. The 1820 treaty was followed by the 1835 treaty which was renewed every year until 1843, when it was extended for ten years. In 1853 a Treaty of Peace in Perpetuity was finally signed 'to be watched over and enforced by the British Government.' After the conclusion of this last treaty the British changed the name of the Omani Coast from the Pirate Coast to the Trucial Coast.[17]

The 1820 treaty specifically aimed at inflaming the feuds among the Arab tribes so as to weaken them further and prevent them from taking a unified stand. To witness this Sanger says:

> The local Sheikhs were forced to sign a general treaty of peace
> with the British in January, 1820, under which the Qawasims
> agreed not to engage in piracy against outsiders, although
> they were allowed to fight each other as much as they wished
> by land and by sea.[18]

The British then did not need to occupy themselves with maintaining order in the interior. In 1820, the oil riches were not known to exist under the barren and unproductive coast of Arabia. Indeed, the coast was 'hell on earth.'[19] It was 'a cardinal principle of British policy in the Gulf not to become involved on the Arabian mainland, and that principle was to hold good until well into the twentieth century.'[20] To be involved would be a burden without benefit.

When, however, news brought out the presence of oil riches under the barren Arabian shores, the area saw the most 'exclusive' treaty arrangements one could think of. Close to the end of the nineteenth century, the presence of oil under the surrounding shores of the Persian Gulf became publicly known. In 1871, a group of German geologists brought home with them encouraging reports on the existence of oil in Iraq. These findings were later confirmed by 'the reports of de Morgan (1892), Stahl (1893), Colonel Maunsell (1897), and Baron von Oppenheim (1899).' The prospects of oil in Persia led Baron Julius de Reuter, the founder of the news agency, to seek a mineral concession there in 1872.[21]

Once again, the news of oil brought to the forefront European rivalry for a position in the Persian Gulf. Toward the end of the nineteenth century Germany, Russia, and France showed increasing interests in the area. The Germans were interested in the construction of a railway from Berlin to Baghdad with its terminal in the town of Kuwait. The Ottoman Railway Company, controlled by the Deutsche Bank, was then formed for this purpose. In 1888, the company secured concessions for the railroad project and mining rights over 20 kilometers along both sides of the railway track. In 1899, the Russians, too, proceeded with their project of constructing their north-south railway across Persia with the terminus on the Persian Gulf.[22]

Britain came alive to confront these new European intrusions. This time the rivalry was about the barren shores of the Gulf, and the British were determined to block. In a series of 'separate but identical treaties' Britain bound every sheikh 'not to cede, mortgage nor otherwise dispose of part of his territories to anyone except the British government, nor to enter into any relationship with a foreign government other than the British without British consent.'[23]

Bahrain signed this treaty in 1880 and in 1892. Bahrain was followed by Oman which signed in 1891. The Trucial Sheikhs signed in 1892. They were followed by Kuwait in 1899. Qatar was then under the nominal suzerainty of the Ottoman Empire, and waited for the Ottoman defeat in World War I to sign the treaty in 1916.[24]

These treaties were further expanded when, in 1911, the British had the sheikhs agree not to grant pearl concessions to foreign powers. A pervasive clause was later (1922) added 'under which the local sheikhs undertook to consult the British Resident regarding any further concessions asked for by other powers.'[25]

In 1903 Lord Curzon visited the Gulf region. In a widely quoted speech, the Lord declared before a gathering of Arab sheikhs:

> We found strife and we have created order. We opened the
> seas to the ships of all nations and enabled their flags to fly in
> peace. We have not seized or held your territory; we are not
> now going to throw away this century of costly and
> triumphant enterprise; we shall not wipe out the most
> unselfish page in history. The peace of these waters must still
> be maintained; your independence will continue to be held;
> and the influence of the British Government must remain
> supreme.[26]

His lordship stated that he found strife and created order, yet blood feuds among the tribesmen of the Omani Coast persisted as late as 1940 and 1948.[27] Recall that the 1820 treaty specifically allowed the Arabs to fight each other as much as they wished. Lord Curzon said he did not seize or hold their territory, yet the sheikhs were chained to such exclusive treaty arrangements that their ramifications are still with us today. He said he was not going to wipe out the most unselfish page in history. Yet it was Kuwait who financed the building of the first two schools in the Sheikhdoms of Sharjah and Ras al-Khaima, and it was Egypt who provided the schoolteachers.[28] Finally, he declared that he would uphold their independence under British occupation. Independence and occupation: a contradiction in terms. This is the British reasoning, however.

This then had been the situation in the lower Gulf region before the granting of oil concessions: fragmentation, weakness and foreign occupation. In the north, in Iraq, the situation was not any different. 'Before World War I,' said Yale William, 'the Persian Gulf area was almost a private British preserve and in the Ottoman province of Mesopotamia the British were predominant, politically and commercially.'[29]

One quick look at the situation in Iran at the time of granting the first oil concession tells us that the country was divided into two spheres of influence; the north under Russia and the south under Britain. The country, according to Sir Arthur Hardinge, the British minister to Persia, was 'ready to be knocked down at once to whatever foreign power bid the highest or threatened most loudly its degenerate and defenseless rulers.'[30] Lord Curzon spoke of the nature of the power struggle that was going on at the time. After his visit to Persia in 1892 he had this to declare:

> I should regard concession by any power of a port upon the Persian Gulf to Russia (that dear dream of so many patriots from the Nova to the Volga) as a deliberate insult to Britain, as a wanton rupture of the status quo, and as an international provocation to war; and I should impeach the British minister guilty of acquiescing in such a surrender as a traitor to his country.[31]

Against this background of weakness, foreign occupation, and the absence of legitimate representatives, oil concessions in the Persian Gulf area were granted. The history of the struggle to obtain these concessions has been, according to Roger R. Sharp, 'very turbulent and marked by international intrigue.'[32] Let us see how.

Iran

It was a British oil hunter, early in the twentieth century (1901), who seriously considered the oil prospects of the Persian Gulf area. William Knox D'Arcy obtained a concession from the Shah of Persia in which he was granted a sixty-year exclusive monopoly for the exploration and exploitation of oil over 500,000 square miles of Persian territory. The concession covered the entire country with the exception of the five northern provinces bordering Russia.

The exclusion of the five northern provinces marked a recognition of the division of the country into two spheres of influence. This division was later formalized in the Anglo-Russian convention of 1907 which had been called to counteract the German rise to power.[33]

D'Arcy's concession was granted over an area co-existing with that of the British sphere of influence. The concession incorporated an item with serious implication for the future development of the northern provinces: it gave the concessionaire exclusive right over the laying of pipelines to the Gulf. Article 6 of the concession states:

> Notwithstanding what is above set forth, the privilege granted by those presents shall not extend to the provinces of Azarabadjan, Ghilan, Mazanderan, Astrabad and Khorassan, but on the express condition that the Persian Imperial Government shall not grant to any other person the right of constructing a pipe-line to the southern rivers of the South Coast of Persia.[34]

With the inclusion of this item, Britain, in effect, rendered the exploitation of the five northern provinces impossible. The measure had its impact on the shaping of the power struggle, as we shall see later.

In 1908 D'Arcy found oil in commercial quantities at Masjid-i-Sulaiman. In 1909 the Anglo-Persian Oil Co. Ltd. (APOC) was formed to operate the concession. On May 20, 1914, during preparation for World War I, the British government bought a controlling interest in APOC. Zuhayr Mikdashi drew a parallel between the motives of the British government in the purchase of shares in the Suez Canal Company in 1875 and the purchase of controlling interest in APOC in 1914. 'In the two cases,' wrote Mikdashi, 'it can be argued, the purchase was prompted by the desire of achieving political advantages – among other benefits.'[35]

Oil discovery in Persia prompted, in 1916, A.F. Khoshtaria, a Russian citizen, to seek and obtain 'with the support of the Czarist government' a seventy-year oil concession over three of the northern five provinces of Persia. The concessionaire was granted an exclusive right for the 'exploration of petroleum and natural gas found in the districts of Ghilan, Mazanderan and Astrabad.'[36] This concession was never acted upon, and might not even be mentioned except its inclusion provides an example of the power struggle that took place at the time.

No sooner had Khoshtaria received his concession than it was abrogated by the Russian revolution of 1917. The revolution brought with it an immediate renunciation of the whole body of treaties, conventions, and concessions concluded by Persia with the Czarist regime. In 1921 this renunciation was further affirmed by the treaty of alliance concluded between the Russian Republic and Persia. An item pertinent to the oil issue was incorporated in Article XIII of the treaty.

> The Persian Government, for its part, promises not to cede to a third power, or its subjects, the concessions and property restored to Persia by virtue of the Present Treaty, and maintain those rights for the Persian Nation.[37]

This Russian moment of idealism, as well as the defeat of Germany in World War I, gave Britain the opportunity to spread its control over the entire Persian territory. Britain, therefore, forced the conclusion of the Anglo-Persian agreement of 1919 which, according to Viscount Grey, the British Ambassador to Washington, amounted to treating 'Persia as a protectorate.'[38]

With regard to the oil question, the British took a number of steps designed to prevent any renewal of Russian claims over the Khoshtaria concession or the fall of that concession into the hands of other powers. They took a stand in support of the Persians in considering the Khoshtaria concession to be invalid for reason of not being ratified by the Majlis (Parliament). In a letter dated December 13, 1919, and addressed to the Prime Minister of Persia, Sir Percy Cox, the British Political Resident in Persia, wrote:

> The British Government prefers to support the standpoint of the Persian Government in that the Khoshtaria concession is invalid. At the same time, I beg to inform your Highness that should the Persian Government desire to grant a new concession in this connection, the British Government hope, in the interest of Persia, that an English Company will be preferred.[39]

To prevent any future quibbling over the Khoshtaria concession and, more specifically, fearing that the concession 'might fall into American hands,' according to the American Ambassador in Paris, the British decided to buy the Khoshtaria concession and form the North Persia Oil Co. Ltd. to put it into operation.[40]

This move gave the British almost complete control over Persia. It was also an instance in which protests were heard from almost every direction: from the Persians, from the Russians, and from the Americans.

The Persians were repelled by the idea of being completely dominated by the British. They maintained that the Khoshtaria concession was not valid by virtue of not being ratified by the Majlis in accordance with Article 23 of the Fundamental Laws of Persia of October 7, 1907. The Persians further pointed out that the concession had been extorted by the Russians during the occupation of Tehran by Russian and British troops in World War I. More importantly for the British involvement in buying the concession, the Persians reminded the British that it was Sir Percy Cox who, in a letter to the Prime Minister, supported the stand taken by the Persians in considering the Khoshtaria concession null and void.[41]

Apparently the protest was not enough to move the British. On November 22, 1921, the Persian Majlis 'unanimously emphasized the invalidity of the Khoshtaria Concession.'[42] On the same date the Persian Majlis granted an American company, Standard Oil of New Jersey, a fifty-year non-transferable and non-assignable concession for oil exploration and exploitation in the five northern provinces.[43]

The granting of the concession to an American concern had been motivated by America's reputation for its distaste for colonization. It was also a move on the part of the Persians to mitigate the British domination. The granting of the concession had, however, been in response to occasional and informal mentionings by the American Chargé in Persia, in accordance with the Department of State instructions, to leading Persians of the 'desirability of American exploitation of North Persia oil.'[44]

The next day (November 23, 1921), the Russian Minister to Persia protested the granting of the concession in the north to an American concern. He brought up the conditions laid down in Article XIII of the 1921 Russo-Persian treaty in support of this protest. Again, the Persians maintained that the said Article was not pertinent since the Khoshtaria concession did not originally exist since it had not being ratified by the Majlis.[45]

As it turned out, this Persian explanation convinced no one – Russians, British or, later, Americans.

Britain objected to the advent of American oil interests in Persia. She instructed her Ambassador to Washington, Sir Auckland Geddes, to present the Department of State with her protestation. Thus, on October 7, 1921, the British Ambassador wrote to the American Secretary of State, Charles E. Hughes, that the Persian government offered 'to an American group an oil concession in North Persia on the ground that the former Russian rights, acquired by a Russian subject named Khoshtaria, has lapsed.' The Ambassador then went on to say that 'His Britannic Majesty's Ambassador is instructed to point out that these rights were taken over in proper form some time ago by a British firm and,' warned the Ambassador, 'that His Majesty's Government have left the Persian Government in no doubt that the British right to the concession is valid and, if questioned, will receive official support.'[46]

The warning was rejected by the US Secretary of State. In a letter dated October 15, 1921, Secretary Hughes wrote the British Ambassador that the fundamental law of Persia requires that all concessions must be approved by the Majlis to be valid. He then reminded the British of the exclusive control they already had over the greater part of Persia. Finally, the Secretary chose to conclude his letter in this manner:

> I feel justified accordingly, in assuming that unless the claims in question could be properly established there would be no purpose on the part of the British Government to employ its influence to prevent the enjoyment by American citizens of such opportunity as remains for acquiring a minor participation in the petroleum industry of Persia.[47]

There ensued a long diplomatic struggle and it took the Americans over 30 years to finally establish a foothold in the petroleum industry of Persia early in the 1950s. In the meantime, while this struggle was going on between the American and British governments, the traders came together to reconcile their differences. Standard Oil of New Jersey realized that an oil concession without the right to transport the oil would remain just an oil concession. For the oil from North Persia to reach its markets, it must first be transported either farther north to the Black Sea through Russian territory, or to the south and the Persian Gulf through an exclusively British domain. Of course there was no point in thinking of pumping the oil to the north and consequently

Standard Oil of New Jersey was, in essence, left with only one choice – to transport the oil to the south through British domain.[48]

Hence, the American Ambassador to Britain, George Harvey, wrote to the US Secretary of State that he was privately told by officials of the British Foreign Office that 'Greenway of Anglo-Persia and Bedford of Standard Oil had met in London recently, and that it was understood they had arrived at informal agreement to operate jointly. This understanding,' said the Ambassador, 'may put a stop to dissensions.'[49]

Simultaneously, representatives of the British Embassy in Washington paid a visit to the US Under-Secretary of State and informed him that 'they had been advised that the Anglo-Persian Oil Company and the Standard Oil Company had come to an agreement whereby their interests in the Persian fields would be pooled.' The representatives stated that the British Ambassador to Washington 'would like to be assured that this met with the approval of the Department.'[50]

As a result, an agreement was reached between the two companies on February 28, 1922, to form a joint venture. Subsequently, the agreement was presented to the Persian government, which rejected it on March 4, 1922. Standard Oil of New Jersey failed to abide by the terms of the concession not to assign or transfer the concession to a third party. It associated itself with 'a British concern in which the British government has a predominant influence.' Such an association, said the Persian Ambassador to Washington, is 'peculiarly distasteful' to his government.[51]

Thereafter, within a period of eighteen months, the attitude of the American government reversed. The American attitude changed from one of decisive support of the Persians' point of view on the Khoshtaria question to one of vagueness or noncommittal at best. On February 28, 1922, the *New York Times* carried a lengthy article on the oil issue in North Persia. It quoted the *National Petroleum News* as saying that the American government recognized the Khoshtaria concession. On March 6, 1922, the Persian Ambassador to Washington presented the article to the

Department of State and requested a denial of the statement of the *National Petroleum News*.[52] On March 14, 1922, the Persian Ambassador received the following from Secretary Hughes:

> The Department's position with reference to claims to concessions in northern Persia has been and is that it is not sufficiently informed at this time to express an opinion on the legal status of any of the contracts to which Khoshtaria may have been a party or which may have been transferred to him.[53]

A number of explanations were offered for the change of the American stand on the Khoshtaria question. One of these was the realization that the British had an exclusive right over the laying of pipelines and, accordingly, there could be no development of the oil in the north without the British acquiescence. More important, however, was the suggestion that the change was prompted by the understanding that ended the struggle between the British and Americans over the oil prospects in Mesopotamia (Iraq). In this context, the authors of *The Resurgent Years* had hinted at this development by saying that 'the strength of the Anglo-Persian group in the Turkish Petroleum Company might have been enough to force the Jersey Company into the ill-fated agreement to cooperate in Persia.'[54] As we shall see later, the Americans got half of the British share in the Turkish Petroleum Company.

For their part, the Persians were determined not to allow the British in. Therefore, on June 11, 1922 the Persian Majlis annulled the concession granted to Standard Oil. At the same time the Majlis empowered the government 'to negotiate a petroleum concession in north Persia with any independent and responsible American company.' Again the Majlis insisted that the granting of concessions should be conditional on their being non-transferable and non-assignable.[55]

Sinclair Consolidated Oil Corporation, an American concern, seized the opportunity to present its offer to the Persian government. At that time Sinclair had a harmonious relationship with the Russians. It had a Russian concession for oil exploitation on the island of Sakhalin. Thus the company had no problem in pumping the oil in northern Iran northward through Russian territory. In 1924 however, Sinclair lost its position with the Russians and, consequently, lost its concession on Sakhalin. Since there was no way of pumping the oil southward through British domain, Sinclair gave up its concession with Persia.[56]

In 1937 the American Oil Company sought the development of North Persia oilfields. Like the previous attempts this one proved of no avail.

Thus far three American attempts had failed before the Americans were able to share the oil resources of Iran with the British. This finally was achieved in 1954 – the discussion of which is reserved for the next chapter.

Iraq

Oil was known to have existed in Iraq since 1871. Although the British had the upper hand in the Ottoman province of Mesopotamia (Iraq), it was the Germans who were the first to interest themselves in the oil development of Mesopotamia. The Germans obtained their concession in 1904 over 20 kilometers on both sides of the then projected Berlin–Baghdad railway. However, before World War I, there had been four rivals seeking oil concessions in Mesopotamia. Besides the Germans (German-Deutsche Bank), there were the British-D'Arcy group (Anglo-Persian Oil Company-APOC); the Dutch-Anglo-Saxon Oil Company, a subsidiary of the Royal Dutch-Shell group; and Rear Admiral Colby Chester, representing an American group sponsored by the New York Chamber of Commerce.[57]

In 1911 these European rivals pooled their differences to establish African and Eastern Concessions Limited – a British concern. Calouste Serkis Gulbenkian, an Armenian, was instrumental in the emergence of this company. Gulbenkian was a member of the managing directors of the National Bank of Turkey, a British institution, established in 1910. He was also a close friend of Henri Deterding of the Dutch-Shell group. Gulbenkian's objective was to arouse in the National Bank of Turkey interest in the oil potentialities of Mesopotamia and to reconcile British and German differences. As a result, African and Eastern Concessions with a capital of £80,000 was formed. Thirty-five percent of the shares were allotted to the National Bank of Turkey and 25 percent to the German-Deutsche Bank and the Royal Dutch-Shell group each. Gulbenkian pocketed the remaining, 15 percent. In 1912 the name of the company was changed to the Turkish Petroleum Company (TPC).[58]

This arrangement was met by the British with little enthusiasm. It left them with less than a controlling share over the company and it ignored the Anglo-Persian Oil Company, a wholly British concern with real experience in the oil business. The National Bank of Turkey, though a British institution, was less likely to be really involved in the oil business and the possibility of selling its share to a non-Britisher, therefore, was not inconceivable.[59]

The British therefore were withholding their support to push the company through the Ottoman authority when, in January 1914, it was brought out in the news that representatives of Standard Oil Company, an American concern, were seeking a concession in Mesopotamia.[60]

This news prompted the British and the Germans to meet in March 1914, in the British Foreign Office to reconcile their differences. The result of the conference was the Foreign Office Agreement, named after the place of the gathering. Under this agreement, Anglo-Persian Oil Company was given 50 percent of Turkish Petroleum Company. Deutsche-Bank got 25 percent and Royal Dutch-Shell another 25 percent. Gulbenkian was given five percent non-voting stock from the British and Dutch interests. The capital of TPC was raised to £160,000.[61]

The Foreign Office Agreement incorporated a self-denying clause which would become the center of a future controversy when it was included in the Red Line Agreement of 1928.

According to this clause, members of TPC agreed not to interest themselves in the oil development of the Turkish Empire independently. Article 10 of the agreement follows:

> The three groups participating in the Turkish Petroleum Company shall give undertakings on their own behalf and on behalf of the companies associated with them not to be interested directly or indirectly in the production or manufacture of crude oil in the Ottoman Empire in Europe and Asia, except in that part which is under the administration of Egyptian Government or of the Sheikh of Kuwait, or in the 'transferred territories' on the Turco-Persian frontier, otherwise than through the Turkish Petroleum Company.[62]

Shortly thereafter, the British and German ambassadors in Constantinople presented the Turkish Grand Vizier (Prime Minister) with a request to grant the TPC a concession for oil exploitation in the provinces of Mosul and Baghdad in Mesopotamia. Said Halim Pasha, the Grand Vizier, consented to the granting of the concession to the company in a letter addressed to the German and British Ambassadors to Constantinople dated June 28, 1914. This letter is important to future controversies, and reads in part:

> The Ministry of Finance being substituted for the Civil List with respect to petroleum resources discovered in the vilayets of Mossoul and Bagdad, consents to lease these to the Turkish Petroleum Company, and reserves to itself the right to determine hereafter its participation, as well as the general conditions of the contract.
>
> It goes without saying that the Society must undertake to indemnify, in case of necessity, third persons who may be interested in the petroleum resources located in these two vilayets.[63]

Before the matter could be pushed further to receive the assent of the Turkish Parliament, war broke out on August 4, 1914. The war, as we know, ended with the victory of the Allies, and in the Peace Conference Britain and France sat down to allot the Arab territories. Territorial distribution was largely entangled with the oil question, but Britain and France finally reached an agreement at San Remo in April 1920. Under this agreement Mesopotamia, Palestine, and Jordan were mandated to Britain. Syria, including what is now Lebanon, was mandated to France.[64]

France, in control of Syria, the only way that Mesopotamian oil could reach the Mediterranean, was allotted, according to Lord Curzon, the German interest in the Turkish Petroleum Company. Britain then kept her 50 percent in the company while Royal Dutch-Shell was endorsed on its 25 percent. Gulberkian was allowed to keep his five percent.[65]

This arrangement brought about the immediate revolt of the Arabs. The British promised them independence and unification in compensation for their support of the Allies' cause. They then divided their country, mandated their territories, and promised part of their land to other people. These actions precipitated the Iraqi revolt of 1920, which, however, 'was quickly crushed by the British-Indian forces but not before nearly 10,000 lives had been lost in the fighting.'[66]

The Americans, too, protested the San Remo agreement. They were worried about the dwindling of their domestic oil supplies, and the control of foreign oil reserves by a British-French trust was thought to be to their detriment. Also, to use the words of Charles Hamilton of the Gulf Oil Corporation, 'America had contributed much in the way of manpower, fuel and material in the World War effort and so should rightfully share in the "spoils."'[67]

Thus, in 1921, representatives of seven American oil companies were called by the then US Secretary of Commerce, Herbert Hoover, to consider the oil prospects in Mesopotamia. The representatives, according to Hamilton, were encouraged to believe that they were furthering the interests of the United States by proceeding to explore foreign countries for oil.[68]

To counteract the discriminatory measure embedded in the San Remo agreement, the United States advocated the principle of an 'open door' policy for equal commercial opportunities to the nationals of all nations in the mandated territories. The foundations of this principle were, however, severely challenged by Lord Curzon in his note of February 28, 1921.

Lord Curzon maintained that the provisions of the San Remo agreement and the admission of France to TPC were just 'an adaptation of pre-war arrangements to existing conditions.'[69]

It appears he also suspected the motivation behind the United States' insistence on having a share in the oil of Mesopotamia was to avoid the exhaustion of domestic oil resources of the United States. He thus said that although future potentialities were problematic, it was 'the undisputed fact that at present United States soil produces 70 percent and American interests in adjoining territory control a further 12 percent of the oil.' Lord Curzon continued:

> It is not easy therefore to justify that the United States Government's insistence that American control should now be extended to sources which may be developed in mandated territories, and that too at the expense of the subjects of another state who have obtained a valid concession from the former Government of those territories.[70]

Finally, Lord Curzon contested as well the avocation of the 'open door' principle in a dramatic comparison. 'I observe, however,' he wrote 'that by Article 1 of the Act of the Philippine Legislature of the 31st of August, 1920, participation in [the] working of all "public lands containing petroleum and other mineral oil and gas" is confined to citizens or corporations of the United States or of the Philippines.' Lord Curzon concluded his observation by saying that this Philippine enactment is 'in contradiction with the general principle enunciated by the United States Government.'[71]

The British, it seems, had based the above note on the assumption that the Americans were not in possession of all the facts. So, the first order before the Americans was to prove the invalidity of the Mesopotamia concession granted by the Turkish Grand Vizier. The American Consulate in Berlin was, therefore, instructed to investigate the issue with the other partner in the TPC, i.e., the Deutsche Bank. The investigation of the American Consul in Berlin reaffirmed his previous findings that 'no concession was ever granted to the Turkish Petroleum Company, and that the British claim, as set up in the note of the British Government dated February 28th, rests solely on the letter of the Grand Vizier dated June 28, 1914.' The Consul further stated that he was informed that matters with regard to the TPC concession 'were rapidly coming to a head when the war broke out.'[72]

With this information in hand, the United States went on to press its demand. The American Ambassador in London was instructed to present Lord Curzon with the letter of the Grand Vizier. Resting on the nature and stage of this communication, the Ambassador was further instructed on November 17, 1921, to state that, 'The relations between the Turkish officials concerned and the Turkish Petroleum Company would appear, therefore, to have been those of negotiations of an agreement in contemplation rather than those of parties to a contract.' Hence, said the American Ambassador in conclusion, 'I am instructed to express again the desire of my Government that the claim of the Turkish Petroleum Company, if it continues to be asserted, should be determined by a suitable arbitration.'[73]

From then on the British voice took a softer tone. They hoped to change the basis of the argument by changing the status of Iraq, through the termination of the mandatory arrangements and the abolition of capitulary rights for foreigners.[74] They therefore put an Arab government in Iraq and required it to issue a new concession. The Iraqi government, however, asked for a share in TPC in accordance with the San Remo Agreement.[75] The British were not intimidated by this Iraqi request. They stopped the issuance of the Iraqi constitution unless the Iraqi government signed the concession.[76] Accordingly, on March 14, 1925, the Iraqi government issued to TPC a 75-year concession over the whole of Iraq despite the fact that the letter of the Grand Vizier limited the concession to the provinces of Mosul and Baghdad.

To this arrangement the American reaction was vigorous and to the point. In a note dated April 20, 1925 the Secretary of State instructed their Embassy in London to present the British Foreign Office with the statement that 'No conclusion will ... be reached with regard to a mandate until the United States government has had an opportunity to express its view.' The American Chargé in Britain was further instructed to state:

> that no arrangements to which it [United States government] is not a party could modify the rights to which it is entitled in Iraq by virtue of the capitulations of the Ottoman Empire, and it believes that, in accordance with the principles which my Government has consistently advocated and which is not believed His Majesty's Government would be disposed to contest, American nationals should be placed on an equal footing with the nationals of any Allied Power with respect to economic and other rights in Iraq.[77]

This, together with what was then rumored, of the possibility of Standard Oil of New Jersey cutting the British market from its oil supply, brought the British to agree to the admission of American companies to their exclusive domain – the Persian Gulf.[78]

Accordingly, an agreement was reached in 1928 among five American companies to form the Near East Development Corporation to represent them in the reorganization of the Turkish Petroleum Company. The Near East Development Corporation was composed of Standard Oil Company (New Jersey), 25 percent; Standard Oil Company of New York (Socony-Vacuum), 25 percent; Gulf Oil Corporation of Pennsylvania, 16-2/3 percent; the Atlantic Refining Company, 16-2/3 percent; and Pan-American Petroleum, 16-2/3 percent.

The reorganization of TPC resulted, mainly to the disadvantage of the British, that the four powers (Americans, British, French, and Dutch) were each to hold an equal share of 23.75 percent (with the remaining five to Gulbenkian). Of the five American companies that got in, three were bought by the remaining two: Standard Oil Company (New Jersey), and Socony-Vacuum Oil Company, Inc. The Turkish Petroleum Company (TPC) meanwhile changed its name in 1929 to Iraq Petroleum Company, Limited (IPC). This name has remained ever since, and the ownership is still the same.

In the Foreign Office Agreement of 1914 referred to earlier the owners of TPC agreed to act as one, and that whatever each shareholder did in the area should be to the benefit of all. But in July 1928, when the Americans were admitted on equal footing with the other three companies, a new agreement was reached. It was first called the 1928 Group Agreement and later the Red Line Agreement. It was so called simply because someone had a red pen and drew a line on a map to encircle the whole Arabian Peninsula (except, oddly enough, the British Protectorate of Kuwait) and the whole of the Near East (Iraq, Syria, Jordan, Palestine, Turkey, etc.) to be considered the area of the group operations and covered by the self-denial clause in the original Foreign Office Agreement. The map was then attached to the agreement, and thus the agreement got its name.[79] Among the provisions of the Red Line Agreement were:

1.   The company was incorporated in Britain and should submit together with its signatories to jurisdiction of the English courts and the agreement be governed by English law.

2.   The participating groups act only through jointly owned operating companies in all matters relating to exploration or to the production of crude oil within the defined area. Thus, any concessions obtained by the Company, or any operating

subsidiary or by any of its shareholders, had to be operated, for the benefit of all the participating groups.

3.   All crude oil produced was to be offered at about its approximate cost to each participant in proportion to its ownership of the Company.[80]

It is obvious that the Red Line Agreement provided for a cartel-type arrangement and was in complete contradiction to the principle of the 'open door' policy enunciated by the US Department of State. Yet, it was the Department of State who justified the participation of American companies in this cartel-type exclusive arrangement. In reply to an inquiry over the restrictive measure of the Red Line Agreement made by the Associate General Counsel of the Standard Oil Company of New Jersey, the Department of State pointed out that 'the arrangements contemplated in view of the special circumstances affecting the situation are consistent with the principles underlying the open door policy of the Government of the United States.'[81]

The Red Line Agreement was, at any rate, just an instrument 'to give the United States a stake in the Middle East.'[82] When, however, its restrictive measures hindered the Americans, they instigated its destruction.

Between 1914 and 1928, the extent of the huge oil reserves in the Persian Gulf area were appreciated to be mainly in Iran and Iraq. By the late 1930s the huge reserves of Saudi Arabia began to be realized. But for the American companies – members of the Red Line Agreement – to invest in Saudi Arabia (who had already awarded a concession to Standard Oil of California which in 1936 brought in Texaco as an equal partner) would have brought in the British, French, and Dutch to an exclusively American concession.

In 1940 Standard Oil of New Jersey approached its partners for renegotiation of the Red Line Agreement. They refused, and Standard Oil of New Jersey took a strong stand. It used the legal argument that since the Germans occupied France in World War II,

70

the French part of the agreement was null and therefore the whole restrictive feature of the agreement was null. This, despite the fact that the corporation was British and under British and not French law. Litigation and negotiations continued while simultaneously Standard Oil of New Jersey and Socony, members of the Red Line Agreement, began to work out the terms of their participation with Standard Oil of California and Texaco, holders of the Saudi Arabian concession. In March 1947, Standard Oil of New Jersey and Socony got 30 and 10 percent, respectively, of the Arabian American Oil Company (Aramco) in Saudi Arabia. In return Standard Oil of New Jersey paid $76.5 million and Socony paid $25.5 million to Standard Oil of California and Texaco, the owners of Aramco.[83]

The French and Gulbenkian sued the American companies. Before the legal proceedings went to trial, however, the French were offered to take more than their share in the oil production of IPC for their acceptance of ending the Red Line Agreement. Similarly, Gulbenkian was offered 3.8 million tons every year in excess of his five- percent share for 14 years. Both then agreed to drop the case, and the Red Line Agreement came to its end in 1948.[84]

Bahrain, Kuwait, and Saudi Arabia

For over two centuries the British stood against the Portuguese, the Dutch, the French, the Russians, and the Germans. The Red Line Agreement of 1928, however, signaled the end of the British domination in the Persian Gulf. Iraq provided the stepping stone for America's involvement in Gulf oil. From that time on the British had to relinquish a larger and larger share to their American partner.

As it was D'Arcy and Khoshtaria in the case of Iran and Gulbenkian in the case of Iraq, it was Major Frank Holmes, a New Zealander, in the case of Bahrain, Kuwait, and Saudi Arabia. Major Holmes formed a company and called it Eastern and General Syndicate – a British concern. He then started looking for oil concessions, and he decided Bahrain should be his first target. Bahrain was under the British exclusive treaty arrangements and, Eastern and General Syndicate, the British establishment, should have had no difficulty in acquiring a concession there. On December 2, 1925, Eastern and General Syndicate got from the Sheikh of Bahrain an oil concession of about 100,000 acres. However, no sooner had Major Holmes pocketed the concession than he took to the roads hunting for any purchaser.[85]

On November 30, 1927, Major Holmes was able to interest Gulf Oil Corporation – an American concern. Gulf Oil, however, was then a member of the Red Line Agreement (the IPC group) and, as such, it was denied the right to take the option over Bahrain by the other European members of the Red Line Agreement.[86]

Gulf Oil then assigned, with the consent of Major Holmes, on December 21, 1928, its option rights to Standard Oil Company of California, a newcomer to the area which was not involved in the restrictive measures of the Red Line Agreement. By then, however, the option was due for renewal. Attempts were made to secure a one-year renewal of the concession when the British were alerted to this new development – the intrusion of a purely American concern into an exclusively British domain. Accordingly, the British Colonial Office made the renewal of the concession:

> … contingent upon the insertion in the original concession
> agreement of a clause providing, among other things, that the
> managing directors and a majority of other directors should
> be British subjects, that the concessionaire company should be
> British-registered, and that none of the rights and privileges
> which the sheikh had granted in the concession should be
> controlled directly or indirectly by foreigners.[87]

The Department of State became aware of this development, and on March 28, 1929, the American Embassy in London was instructed to request the British government to make a statement of policy on granting concessions in areas such as Bahrain.[88]

On May 30, 1929, the British Foreign Office agreed in principle to the participation of United States interests in the Bahrain concession. Negotiations were begun between the American company and the British Colonial Office. Accordingly, Standard Oil of California formed a wholly owned subsidiary, Bahrain Petroleum Co. Ltd., which was registered under Canadian law. 'Here again,' stated the Department of State, 'the prompt and positive action by the State Department had secured results favorable to an American-owned company.'[09] The Department of State went on to say:

> By securing the entry of American oil interests into Bahrain, the way was paved for some American interests to obtain concession in nearby Arabia.[90]

Oil was discovered in Bahrain in 1932. Its discovery was the cause of great disturbances to the European members of IPC, and Anglo-Persian in particular. The discovery was regarded as a threat to their interests.[91] Anglo-Persia wanted to contain the problem before it got out of hand, and tried to win a concession over the 70,000 acres in Bahrain which were not covered by the Standard Oil of California concession. The American partners in the Red Line Agreement, however, objected to this move on the part of Anglo-Persia. They pointed to the restrictive measures of the Red Line Agreement which at an earlier date barred Gulf Oil Corporation from obtaining a similar option in Bahrain.[92]

The British apparently drew a lesson from this Bahrain development. They witnessed the Americans involve in their exclusive domain a company which was not a member of the Red Line Agreement and yet able to get away with the concession solely for itself. Thus when a similar move was made by Gulf Oil Corporation in Kuwait, another British preserve, the British decided this should not happen.

Gulf Oil Corporation, a signatory to the Red Line Agreement, sold its shares in IPC to the other American members of the Near East Development Corporation. Now unhampered by the restrictive measures of the Red Line Agreement, Gulf Oil tried in 1929 to seek an oil concession in Kuwait. Like Bahrain, Kuwait was under the British exclusive treaty arrangements, but was not part of the Red Line area.

The arrival of Gulf Oil into Kuwait was, however, blocked by the British Colonial Office which insisted on the application of the 'nationality clause.' According to this clause no one but a British subject or firm could obtain a concession in Kuwait.[93]

Again, the Department of State was notified and on December 3, 1931, the American Embassy in London was instructed to make representation with the view to securing equal treatment for American firms. Four years, however, were to elapse before the British consented to American participation in the development of Kuwait oil resources. On November 2, 1932, the American Ambassador in London wrote that the 'delay in reaching a settlement in the matter of the Kuwait oil concession was becoming exasperating.'[94]

Obviously, the British were determined not to let the Americans develop the oil resources of Kuwait solely by themselves. Before they committed themselves to the question of equal participation with the Americans, however, the British wanted to make sure that Kuwait was worth the investment. Thus, while they were negotiating the American participation, twice they sent their geologists to Kuwait to study the surface geology of the ground. As a result, the American Embassy in London was instructed to request the British Foreign Office to prevent Anglo-Persia from proceeding with its geological exploration until a decision was reached over the participation of American companies in Kuwait.[95]

It appeared that the results of these geological surveys were encouraging and, accordingly, the British agreed, on April 9, 1932, on the participation of Gulf Oil; not, however, before they induced Anglo-Persia to submit a counter offer.[96] However, on the British government's recommendation, both offers were rejected by the Sheikh of Kuwait on January 9, 1933.[97]

Shortly thereafter, both companies got together to reconcile their differences and form the Kuwait Oil Company Ltd., a British concern equally owned by its two parents, Anglo-Persia and Gulf Oil Corporation. Kuwait Oil Company presented its offer to the Sheikh of Kuwait who, on December 23, 1934, granted the company a 75-year concession over Kuwait, its islands and territorial waters. In May 1938, the largest known oil pool in the world was discovered – the Burgan field.[98]

Saudi Arabia was encircled in the Red Line area. The King of Saudi Arabia, however, 'had a lively distrust of the British.' Thus when the King was approached by Major Frank Holmes and Anglo-Persia, both British entities, but with Anglo-Persia controlled by the British, the King was in favor of Major Holmes. On May 6, 1923, the King granted Major Holmes an exclusive option for oil and mining rights over Al-Hasa province. A year later Holmes was able to secure another option over the undivided half of the Saudi–Kuwaiti Neutral Zone. Nevertheless, these options were invalidated by reason of Major Holmes' 'failure to make payments or to initiate any exploration work program.'[99]

The news of the oil discovery in Bahrain soon aroused interest in the mainland. In 1933 both Standard Oil of California – an American company – and Iraq Petroleum Company – a British concern – approached the authority in Saudi Arabia for an oil concession. A sixty-year concession over Al-Hasa province, a total area of 360,000 square miles – one of the largest oil concessions in the world – was granted to the American company. According to Charles Rayan, petroleum adviser to the Department of State, 'the total absence of any pressure on the part of the American Government was one of the deciding reasons for the award of the concession to an American company.'[100]

Oil was found at the Dammam field in 1936, and commercial production followed in 1938. In December 1936, the Texas Oil Company, obviously encouraged by the oil discovery in Saudi Arabia, acquired a 50 percent share in the Standard Oil of California interest in Saudi Arabia. The merger of both interests gave rise to the new company controlling the oil in Saudi Arabia, known as the Saudi Arabian American Oil Company (Aramco). For the acquisition of this interest the Texas Oil Company paid $3 million in cash and $18 million in deferred payment to be settled out of the oil produced from the Saudi Arabian fields.[101]

The oil discovery in Saudi Arabia had, of course, encouraged a number of interested parties to look for oil concessions in the area then uncovered by Aramco's concession. Saudi Arabia awarded the concession to Aramco even though this company offered, according to Rayan, 'less than the Government-controlled Japanese and German companies, whose diplomats at Jiddah were extremely pressing with their offers.' The original area of Aramco was increased by 80,000 square miles, thereby making Aramco's total concession an area about 440,000 square miles – almost two-thirds of the country.[102] To look for a reason for this move on the part of Saudi Arabia, one cannot see other than the reputation that preceded America in the whole Arab world as a force of liberation with a distinct distaste for colonization.

As noted in the discussion about the Red Line Agreement, Aramco agreed in 1947 to the participation of Standard Oil of New Jersey and Socony-Vacuum in its capital stock. Accordingly, Standard Oil of New Jersey and Socony-Vacuum received 30 percent and 10 percent, respectively, of the shares of Aramco. Standard Oil of New Jersey then made a payment of $76.5 million and Socony $25.5 million for their respective shares. Thus two American companies, members of the Red Line Agreement, were added to an oil concession which was within the Red Line area, yet the addition created an exclusively American trust.

The British sprang into action to blockade this new American tactical exclusion. In 1935, Anglo-Iranian (formerly Anglo-Persia) obtained a 75-year oil concession over the Sheikhdom of Qatar. Qatar was under the British exclusive treaty arrangements, and she was also part of the Red Line area. Anglo-Iranian, therefore, assigned the concession to Iraq Petroleum Company, the founder of the Red Line Agreement which equally represents all the powers of the day.

More important for equal involvement and participation of all the powers were the activities embarked upon by IPC in getting concessions in almost every place within the Red Line circle. In the period between 1933 and 1945, IPC obtained licenses and concessions in Syria, Lebanon, Palestine, Jordan, western Saudi Arabia, Yemen, Aden Protectorate, Oman and the Trucial Coast. Most of these areas appeared to be dry and they were surrendered by IPC, with the exception of Oman and the Trucial Coast whose oil resources were developed late in the 1950s.[103]

Endnotes

[1] The *Middle East and North Africa, 1970-71* (17th edn.; London: Europa Publications Limited, 1970), 557 and 586.

[2] J.B. Kelly, *Eastern Arabian Frontiers* (London: Faber and Faber, 1964), 57-58; *The Middle East and North Africa*, 1969-70 (16th edn.; London: Europa Publications Limited, 1969), 250.

[3] Kelly, op. cit., 57-58; *The Middle East and North Africa, 1969-70*, op. cit., 409.

[4] J.F. Standish, 'Britain In the Persian Gulf,' *Contemporary Review*, CCXI, No. 1222 (November, 1967), 235; Richard Coke, *The Arab's Place in the Sun* (London: Thornton Butterwoth Ltd., 1929), 293-294.

[5] *The Middle East and North Africa, 1969-70*, op. cit., 565.

[6] Richard H. Sanger, *The Arabian Peninsula* (Ithaca, New York: Cornell University Press, 1954), 171.

[7] Kelly, op. cit., 48; *The Middle East and North Africa, 1970-71*, op. cit., 586.

[8] Coke, op. cit., 298; Norman C. Walpole, *Area Handbook for Saudi Arabia* (Washington DC: Government Printing Office, 1966), 3.

[9] Kelly, op. cit., 48; *The Middle East and North Africa, 1969-70*, op. cit., 21; Sanger, op. cit., 171.

[10] Sanger, ibid., 171-172.

[11] Ibid.

[12] Kelly, op. cit., 58; Sanger, op. cit., 172.

[13] Sanger, ibid.

[14] Sir Rupert Hay, *The Persian Gulf States* (Washington DC: The Middle East Institute, 1959), 145.

[15] Ibid., 146.

[16] Ravinder Kumar, *India and the Persian Gulf Region 1858-1907: A Study in British Imperial Policy* (New York: Asia Publishing House, 1965), 211.

[17] Kelly, op. cit., 61-62; Sanger, op. cit., 172.

[18] Sanger, ibid., 172.

[19] Roderic Owen, *The Golden Bubble: Arabian Gulf Documentary* (London and Glasgow: Collins Clear-Type Press, 1957), 165.

[20] Kelly, op. cit., 75-76.

[21] Charles W. Hamilton, *Americans and Oil in the Middle East* (Houston, Texas: Gulf Publishing Company, 1962), 81; Christopher Tugendhat, *Oil: The Biggest Business* (New York: G.P. Putnam's Sons, 1968), 59.

[22] *The Middle East and North Africa, 1970-71*, op. cit., 586; Hamilton, op. cit., 81; Tugendhat, op. cit., 59.

[23] *The Middle East and North Africa, 1970-71*, op. cit., 586.

[24] Ibid., 557, 574, 581 and 586; *The Middle East and North Africa, 1969-70*, op. cit., 409; William Luce, 'Britain in the Persian Gulf, Mistaken Timing over Aden,' *Round Table*, LVII, (July, 1967), 279.

[25] Sanger, op. cit., 173.

[26] Quoted in Standish, op. cit., 239.

[27] Sanger, op. cit., 182.

[28] Owen, op. cit., 165.

[29] Quoted in Hamilton, op. cit., 317.

[30] Sir Arthur H. Hardinge, *A Diplomatist in the East* (London: J. Cape Limited, 1928), 280.

[31] Quoted in Olaf Caroe, *Wells of Power, the Oilfields of South-Western Asia, A Regional and Global Study* (London: Macmillan and Co., Ltd, 1951), 64-65.

[32] Roger R. Sharp, 'America's Stake in World Petroleum,' *Harvard Business Review*, XXVIII, No. 5 (September, 1950), 29.

[33] Caroe, op. cit., 68-69.

[34] League of Nations, *Official Journal*, 13th Year, XIII^2, 1932, 2306.

[35] Zuhayr Mikdashi, *A Financial Analysis of Middle Eastern Oil Concessions: 1901-65* (New York, Washington DC and London: Frederick A. Praeger, Publishers, 1965), 16.

[36] Ibid., 28; United States, Department of State, *Papers Relating to the Foreign Relations of the United States 1920*, III (Washington DC, 1936), 351-352.

[37] League of Nations, *Treaty Series, 1922*, IX, 407.

[38] Great Britain, Foreign Office, *Documents on British Foreign Policy 1919-1939*, E.L. Woodward and Rohan Butler (eds.), IV, 1st Series (London, 1952), 1214.

[39] US, Congress, Senate, *Oil Concession in Foreign Countries*, Doc. No. 97, 68th Cong., lst Sess. (Washington DC: Government Printing office, 1924), 106-107.

[40] See Mikdashi, op. cit., 199; see also a letter dated October 10, 1919 sent by Viscount Grey, the British Ambassador to Washington, to Earl Curzon in which the Ambassador expressed the American feelings towards the Anglo-Persian agreement of 1919 as designed 'to exclude American commerce or influence.' *Documents on British Foreign Policy 1919-1939*, op. cit., 1200.

[41] *Oil Concession in Foreign Countries*, op. cit., 94-95.

[42] Ibid., 96.

[43] Ibid., 120; United States, Department of State, *Papers Relating to the Foreign Relations of the United States, 1921*, II (Washington DC, 1936), 648.

[44] *Papers Relating to the Foreign Relations of the United States, 1921*, ibid., 643. The American Chargé in Persia wrote on November 22, 1921, to the Secretary of State asking for Standard Oil of New Jersey to send its representatives to take over since 'the legation has already gone as far as it properly can in pushing the matter through.' Ibid., 648.

[45] *Oil Concessions in Foreign Countries*, op. cit., 94.

[46] *Papers Relating to the Foreign Relations of the United States, 1921*, op. cit., 643-644.

[47] Ibid., 645-646.

[48] See Mikdashi, op. cit., 29.

[49] *Papers Relating to the Foreign Relations of the United States, 1921*, op. cit., 652-653.

[50] Ibid., 654.

[51] *Oil Concession in Foreign Countries*, op. cit., 105, 115 and 120.

[52] Ibid., 101 and 103.

[53] Ibid., 104.

[54] G.S. Gibb and E.H. Knowlton, *The Resurgent Years, History of Standard Oil Company (New Jersey) 1911-1927* (New Jersey: Harper & Bros., 1956), 315; Benjamin Shwadran, *The Middle East, Oil and the Great Powers* (New York: Frederick A. Praeger, 1955), 90; Mikdashi, op. cit., 29.

[55] *Oil Concession in Foreign Countries*, op. cit., 113; United States, Department of State, *Papers Relating to the Foreign Relations of the United States, 1923*, II (Washington DC, 1938), 724.

[56] Shwadran, op. cit., 94.

[57] Stephen Hemsley Longrigg, *Oil in the Middle East: Its Discovery and Development* (2nd edn.; London: Oxford University Press, 1961), 28-29; Shwadran, op. cit., 194; Mikdashi, op. cit., 65.

[58] Longrigg, op. cit., 29-30; Tugendhat, op. cit., 68-69.

[59] David H. Finnie, *The Middle East Oil Industry in Its Local Environment* (Cambridge: Harvard University Press, 1958), 30; Longrigg, op. cit., 30; Tugendhat, op. cit., 69.

[60] Finnie, op. cit., 30; see also Lord Curzon's note of February 28, 1921, in *Papers Relating to the Foreign Relations of the United States 1921*, 81-82.

[61] United States, Department of State, *Papers Relating to the Foreign Relations of the United States 1927*, 11 (Washington DC, 1942), 821-822.

[62] Ibid., 822.

[63] *Papers Relating to the Foreign Relations of the United States, 1921*, op. cit., 90.

[64] Joseph E. Pogue, 'Must An Oil War Follow This War?' *Atlantic Monthly*, CLXXIII, No. 3 (March, 1944), 42.

[65] *Papers Relating to the Foreign Relations of the United States, 1921*, op. cit., 83.

[66] Hamilton, op. cit., 79.

[67] Ibid., 83.

[68] Ibid.; Standard Oil Company (New Jersey), *A Background Memorandum on Company Policy and Actions* (New York: Standard Oil Company, 1947), 4-5; US, Congress, Senate, Special Committee Investigating Petroleum Resources, *American Petroleum Interests in Foreign Countries, Hearings*, 79th Cong., 1st Sess., June 27 and 28, 1945 (Washington DC: Government Printing Office, 1946), 55.

[69] *Papers Relating to the Foreign Relations of the United States, 1921*, op. cit., 81.

[70] Ibid., 82-83.

[71] Ibid., 84.

[72] Ibid., 85-86.

[73] Ibid., 89-92.

[74] United States, Department of State, *Paper Relating to the Foreign Relations of the United States, 1925*, 11 (Washington DC, 1940), 236.

[75] See Mikdashi, op. cit., 68.

[76] *The Middle East and North Africa, 1970-71*, op. cit., 64-65.

[77] *Papers Relating to the Foreign Relations of the United States, 1925*, op. cit., 236.

[78] See Tugendhat, op. cit., 76.

[79] Standard Oil Company (New Jersey), op. cit., 6.

[80] United States, Federal Trade Commission, *The International Petroleum Cartel, Staff Report Submitted to the Subcommittee on Monopoly of the Select Committee on Small Business*, 82nd Cong., 2nd Sess., No. 6 (Washington DC: Government Printing Office, 1952), 65-67.

[81] *Papers Relating to the Foreign Relations of the United States, 1927*, op. cit., 824.

[82] Tugendhat, op. cit., 88.

[83] *The International Petroleum Cartel, Staff Report submitted to the Subcommittee on Monopoly of the Select Committee on Small Business*, op. cit., 101-104.

[84] Ibid., 105-106.

[85] Ibid., 72.

[86] Ibid., 73.

[87] United States, Department of State, *Papers Relating to the Foreign Relations of the United States, 1929*, 111 (Washington DC, 1944). 90-81.

[88] Ibid., 81; *American Petroleum Interests in Foreign Countries, Hearings*, op. cit., 24.

[89] *American Petroleum Interests in Foreign Countries, Hearings*, ibid.

[90] Ibid.

[91] *The International Petroleum Cartel, Staff Report submitted to the Subcommittee on Monopoly of the Select Committee on Small Business*, op. cit., 73.

[92] Ibid.

[93] *American Petroleum Interests in Foreign Countries, Hearings*, op. cit., 24.

[94] United States, Department of State, *Papers Relating to the Foreign Relations of the United States*, 1932, 11 (Washington DC, 1947), 20.

[95] Ibid., 13 and 16.

[96] Ibid., 14-17.

[97] Mikdashi, op. cit., 81.

[98] Hamilton, op. cit., 190-194.

[99] Tugendhat, op. cit., 89-92; Hamilton, op. cit., 118 and 143; Mikdashi, op. cit., 79.

[100] *American Petroleum Interests in Foreign Countries, Hearings*, op. cit., 24.

[101] *The International Petroleum Cartel, Staff Report Submitted to the Subcommittee on Monopoly of the Select Committee on Small Business,* op. cit., 116.

[102] *American Petroleum Interests in Foreign Countries, Hearings,* op. cit., 24.

[103] Hamilton, op. cit., 101.

The Iranian Crisis: Another Version of the Power Struggle for the Control of Oil

Iran went bankrupt. In 1954 the country was technically insolvent as a result of the major powers' struggle over a share in the control of the Iranian oil. That power struggle, in 1951, led to the agitation for nationalization as a result of which Iran came to sign an agreement not very much different from those signed by Saudi Arabia, Iraq, and Kuwait but without the turmoil.

Nationalization of the oil industry in 1951, however, was not an abrupt movement without its age-long deep grievances. It was the result of a lust carried over from the 'ruthlessness of Victorian expansionism in "backward" areas of the world.'[1] It was also the result of foreign occupation during World War II. And, finally, it was the result of foreign instigation in an attempt to break an old trust for the building of a new one in its place.

The D'Arcy concession, as noted earlier, was granted over an area of 500,000 square miles for an annual payment of 16 percent of future net profit. This concession was controlled by the British government through the control of Anglo-Persia (Anglo-Iranian).

During the darkest days of the great depression, royalty payments made to the Iranian government tended to fluctuate adversely. This is a natural phenomenon, but while the Anglo-Persian net profit dropped from £3.8 million in 1930 to £2.4 million in 1931 – less than 37 percent – the Iranian royalty for the same period went down from £1.29 million to £307,000 – over 75 percent.[2]

This was an instance where the Persians presented their grievances to the British authority. They complained of the existence of a 30-year secret contract between Anglo-Persia and the British Admiralty for the sale to the Admiralty of large quantities of fuel oil under market prices.[3] The Persians further complained that they were barred from investigating the accounting practices of the company, its expenditures, and its profits from its subsidiaries – all having impact on the royalties. They maintained that the company secured a concession on almost all of Persia, yet confined its operations to only a small section of the area under concession. In effect then, they charged, the company was just shutting out other competitors and using the money reaped from Persian fields for oil exploration all over the world. Finally, the Persians had this pervasive complaint to summarize their grievances.

> Not only was the D'Arcy Concession incompatible itself with the interest of Persia, but the concession was granted at a time when the interests and welfare of the country were unfortunately not taken into consideration in the drawing up contracts of this kind, and when those who wished to obtain them took great advantage of the ignorance of the authorities in charge.[4]

In view of these grievances, the Persian government introduced in 1930 a four percent income tax. Anglo-Persia, however, refused to abide by what was not included in the original contract. The government then sought a revision of the D'Arcy concession. But, on 7 August 1931, Sir John Cadman, Chairman of Anglo-Persia, 'categorically stated in a letter to the [Persian] minister of the Court, that the revision of the concession could no longer be contemplated.'[5]

Thereafter, the Persian government, in 1932, asked the company to send its representatives to Tehran in order to arrive at a final agreement. The company, however, saw no 'justification in incurring the expense of sending its experts to Teheran and suggested that a Persian representative be sent to London.' This was more than the sovereign could tolerate, and on November 27, 1932, the Persian Minister of Finance handed B.R. Jackson, Anglo-Persia's representative in Tehran, a cancellation of the D'Arcy concession.[6]

While the League of Nations was discussing the problem, British warships appeared in the Persian Gulf.[7] Nevertheless, a revision of the D'Arcy concession was reached in 1933 – a revision which set the stage for another crisis in 1951.

In accordance with a complicated formula, the Persian government was to receive a tonnage royalty of four shillings for every ton of oil. A guarantee was introduced that this tonnage royalty should not go below a minimum of £750,000. This, however, cannot be considered much of a guarantee, since the guaranteed payment could be achieved with a minimum production of 3.7 million tons. Production had been maintained at a rate of 5.5 to six million tons per year during the five years prior to the revision of the concession.[8]

In addition, the company agreed to pay the Persian government 20 percent of the dividends distributed to ordinary shareholders in excess of £671,250. Why the ceiling of £671,250? The company, a British government enterprise, conceived of a future abstention of dividend payments to the Persian government through the withholding of dividend distribution to its major shareholder – the British government. Thus, while 'His Majesty's Government,' to use Sir Anthony Eden's expression, 'were getting a good rake-off from taxation,'[9] the formula made it possible for the company to continue paying dividends to ordinary shareholders of British subjects without, at the same time, being compelled to make payment to the Persian government. Indeed, this is what had happened in 1949. According to Sir Anthony Eden, then in the Opposition, 'the company was earning 150 percent, or thereabout, but they were still paying 30 percent.'[10] Again, who would determine the amount of profits on the basis of which dividends are calculated, especially if the Persians were barred from investigating the books of the company? As we shall see later, the accumulation of profits in the retained earnings became one of the immediate factors in triggering the crisis of 1951.

Charles Hamilton of Gulf Oil Corporation was puzzled with the terms of the new contract: 'The obvious gain to Persia under the terms of the new concession was the reduction in size of the concession area – from 500,000 square miles to 100,000 square miles. Less obvious was the financial gain to Persia derived from the new royalty basis.'[11]

Even the reduction of the concession area cannot be considered much of a gain since it took the company five years to select its 100,000 square miles. By then the company certainly knew the oil potentialities of every lot.[12]

These were the so-called benefits accrued to the Persians from the conclusion of the 1933 revision of the D'Arcy concession.

The most important achievement the British secured was the prolongation of the D'Arcy concession for an additional 32 years.[13] The original D'Arcy concession was set to expire in 1961; the 1933 revision extended its life to 1993. Financially, while the company was subjected to British taxation, the 1933 revision specifically exempted it from Persian taxation. This difference had been the cause of the great gap between the payments made to both governments – a situation which further fueled the 1951 crisis. In the three years that preceded the crisis, for example, Anglo-Iranian paid in taxation an aggregate amount of over £100 million to the British Treasury. Against this amount Iran received not more than £38.7 million in royalties. In fact, the amount paid to the British government in these three years alone was equivalent to the total royalties paid to the Persian government for the whole period of the company's operations until the date of the crisis.[14]

Though the financial factor was very important in triggering the crisis of 1951, it was by no means the only one. This, however, had been the financial arrangement when the British and Russian troops occupied Iran in World War II on August 25, 1941. Although there had been some Iranian resistance, it was simply overwhelmed, and the Shah was forced to abdicate on September 16, 1941. Thus Sir Winston Churchill, then Prime Minister, could declare in the House of Commons:

> With hardly any loss of life, with surprising rapidity and in
> close concert with our Russian Ally, we have rooted out the
> malignant elements in Teheran; we have chased a dictator
> into exile, and installed a constitutional Sovereign pledged to
> a whole catalogue of long-delayed sorely-needed reforms

and reparations; and we hope soon to present to the House a
new and loyal alliance made by Great Britain and Russia
with the ancient Persian State and people ... [15]

Accordingly, a treaty of alliance with Iran was concluded
on January 29, 1942.[16] Three articles of this treaty are pertinent to
the oil question. The first provided for respect of 'the territorial
integrity, the sovereignty and political independence of Iran.'
Article IV below throws light on the power struggle over the
control of oil during the war. The last paragraph reads:

A special agreement shall be concluded between the Allied
Governments and the imperial Iranian Government defining
the conditions of any transfers to the Imperial Iranian
Government after the war of building and other
improvements affected by the Allied Powers on Iranian
territory.[17]

Article V fixed the date of Allied forces withdrawal 'six months
after all hostilities between the Allied Powers and Germany and
her Associates have been suspended by the conclusion of an
Armistice or Armistices, or the conclusion of peace between them,
whichever date is earlier.'[18]

This treaty was further expounded upon by the conclusion
of the Tehran Declaration of December 1, 1943. The Tehran
Declaration, signed by President Roosevelt, Prime Minister Stalin,
and Prime Minister Churchill, guaranteed the independence,
sovereignty, and territorial integrity of Iran.[19]

Despite the harmonious relationship that prevailed among
the Allies in their dealings with Iran, it was obvious, according to
US Secretary of State Edward R. Stettinius, that 'Iran is perhaps the
most prominent area of the world where inter-Allied friction might
arise.'[20]

Iran had been declared eligible for the Lend-Lease
assistance on March 11, 1942. At first, United States assistance was
provided under the British banner. Very soon, however, it was
obvious that the United States was chartering for it itself 'an

independent and long range policy in Iran.'[21] Three steps were taken by the United States in accentuation of this policy. These steps also signaled American–British conflict over Iran.

First, two American military missions were sent to Iran early in 1942. These missions were instructed to present their military supply requirements directly to Washington despite the British insistence on seeking their view on the matter since Iran was part of their sphere of influence.[22]

Second, the Department of State was alerted to the provision of Article IV of the treaty of alliance cited above. According to the State Department interpretation, the Article envisaged a British and Soviet disposition of 'property which would include American lend-lease materials and affecting the long-range interests of the United States in Iran.' The American ambassadors in Britain and Russia were, therefore, instructed on January 26, 1942, to seek assurance from the British and the Soviets that no action would be taken that might affect American property, rights, or interest in Iran before the United States was consulted.[23]

The relevancy of Article IV and the United States protest against it is seen clearly in the projected construction of 750 miles of oil pipeline in Iraq and Iran, the third factor which signaled United States conflict with Britain over Iran.

The pipeline was first discussed by the joint Anglo-American Mission to Russia. The project then moved to Washington and was presented by Brigadier W.E.R. Blood of the British Staff Mission in Washington who proposed the use of American lend-lease materials and American engineers for the construction of the pipeline 'to parallel the existing Anglo-Iranian Oil Company line.' Later the British proposed the construction of the pipeline by Anglo-Iranian Oil Company using British engineers but American lend-lease materials.[24]

Obviously the British were envisaging a future commercial use of the pipeline. Again, the Department of State was alerted and a meeting was called in January 1942. In addition to officials from the Department of State, present were representatives of the British Home Office of the Iranian Mission, a representative of the Lend-Lease Administration, and officials of private organizations: the

Vice President of Standard-Vacuum Oil Company, a Vice President of Standard Oil Company of California, the President of California-Arabian Oil Company (later Aramco), and a Vice President of the Texas Oil Company.[25]

In the meeting, Wallace Murray, Chief of the State Department's Division of Near Eastern Affairs and later US Ambassador to Iran, presented the conferees with an historical background on how Americans were excluded from participating in the development of oil in the Persian Gulf area and how, in the end, they were allowed a minority share in the IPC group. Considerations were given to the practical uses of the pipeline, whether for military purposes or future commercial undertakings, and how pipeline construction would facilitate American participation in the future oil development of the area. The Anglo-Iranian exclusive right over the laying of pipelines to the south of Iran, had, in the past, barred all non-Britishers from developing the northern provinces; therefore, one could understand the motive behind the conferees' consideration to 'relieve Iran from possible British postwar pressure to make the lines available exclusively to the AIOC.'[26]

In the end the project was thwarted. The episode, nevertheless, again signaled the ensuing power struggle over the control of oil in the Persian Gulf area. It is rumored that at the Tehran Conference of December 1, 1943, the Big Three discussed the postwar development of oil in the Middle East but could not reach an agreement.[27]

Thus, during the war representatives of the oil companies were shuttling back and forth to Iran. Representatives of the Anglo-Dutch Shell group were in Iran seeking an oil concession in the southeast of Persia, and they were followed by representatives of Anglo-Iranian Oil Company. This development prompted the US Secretary of State Cordell Hull to remark that the 'British already had extensive oil concessions in Iran.' Secretary Hull then approved, on November 15, 1943, a move by Standard Vacuum Oil Company, a subsidiary of Standard Oil Company of New Jersey and Socony-Vacuum Oil Company 'to enter into an arrangement with the Iranian government to produce petroleum in Iran.'[28] Early in 1944 representatives of Standard-Vacuum appeared in Tehran only to find themselves preceded by representatives of Sinclair Consolidated Oil Company, an American concern, competing with them and with the British.[29]

At about this time, April 1944, the Iranian government sought the advice of its two American oil experts, Herbert Hoover, Jr and A.A. Curtice, to help it draft a standard oil concession contract. It is at this juncture, it is said, that the American oil experts communicated to the Iranians the news of the 50-50 profit sharing arrangement which was under negotiation between the Venezuelan government and its American concessionaires.[30]

While this American-British maneuver was taking place in the field, the two governments got together in Washington on August 8, 1944, to charter an oil agreement. The gist of this agreement has two aspects: the first would satisfy the United States demand for the application of the 'open door' principle, and the other would satisfy Britain in its demand for respect by each country, and the nationals thereof, of valid concession contracts and the abstention of each country from making any efforts to interfere directly or indirectly with such contracts.[31]

The Russians, already occupying the north, were eyeing this western arrangement with suspicion. They decided, in September 1944, to send a delegation under Deputy Foreign Commissar, Sergei I. Kavtaradze, with a request for an oil concession over the five northern provinces. Kavtaradze's reasoning for the request was that 'if the Persian government chose to grant a concession to the Americans in southeast Persia they should give the Soviets a concession in the north.[32]

This new development prompted the Prime Minister of Iran, Mohammed Saed, on October 9, 1944, to inform United States Ambassador to Tehran, Leland Morris, of the intention of the Iranian government 'to postpone oil negotiations until after the war.'[33] It was at this juncture that a minority leader in the Majlis, Dr Mohammad Musaddeq, sponsored a bill to prohibit the negotiation of oil concessions by any member of the Cabinet without prior approval from the Majlis. On December 2, 1944, that bill passed by a large majority.[34] The British supported the Iranian decision.

Kavtaradze, however, considered the decision to be aimed at 'a rejection of the Soviet proposal and would darken the relations between the two countries.'[35]

In all these major power maneuverings, Dr Herbert Feis, appointed by Secretary Hull as Chairman of the Committee on International Petroleum Policy, reflected on power relations and their impact on the international oil control. Immediately after the settlement of the Iranian crisis and in direct reference to the period around 1944, he said:

> We were worried over the chance that the oil reserves within the United States were about to dwindle while the demand upon them would increase. Important changes in the international situation were forecast. The balance of power between nations was about to become distinctly different than before the war. The spirit of nationalism and the wish for independence among many different countries and peoples were becoming stronger. The political control of parts of the world in which huge oil deposits were located (especially in the Middle East and Far East) was about to be freshly decided.[36]

Next came the issue of the Allied troop withdrawals. In accordance with the terms of the treaty of alliance, foreign troops should be out of Iran by March 2, 1946 – six months after the signing of the armistice with Japan. The United States had all its troops out of Iran by the 1st of January, 1946. Britain agreed to withdraw its troops on the established date – March 2, 1946. The Soviet Union had to be brought before the Security Council of the United Nations twice before they agreed on a date for withdrawing their forces.

By the end of 1945 Iranian forces were prevented by the Soviet troops from entering the province of Azerbaijan, and in mid-December a separatist movement was encouraged by the Soviets to declare the rise of the 'Autonomous Republic of Azerbaijan.' In January 1946, Iran brought the case before the Security Council which on January 30, 1946, directed both Iran and the Soviet Union to solve their problem through direct negotiation.[37]

Apparently in a move to appease the Soviets and win any concession from them on the question of troop withdrawal, the Prime Minister of Iran, Ahmad Qavam as-Saltaneh, decided to visit Moscow (February 19–March 11, 1946) at the head of a large Iranian delegation. While in Moscow the Premier was presented with a list of Soviet demands:

1. USSR troops to remain in some parts of Iran for an indefinite period.
2. The Iranian Government to recognize the internal autonomy of Azerbaijan.
3. The formation of an Iranian-USSR joint-stock company in which fifty-one percent of the shares would be owned by the USSR and forty-nine percent by Iran for the purpose of developing and exploiting the oil of the North.[38]

The Soviets thus spelled out their intention and on March 26, 1946, the Security Council once again took up the issue at the Iranian request. Before the Council was to convene, however, the Soviet Union announced its decision to withdraw its troops in a period of five or six weeks if nothing occurred. Accordingly, the Council approved on April 4, 1946, a resolution introduced by the United States to postpone further the discussion of the Iranian case until May 6-six weeks from the Soviet announcement.[39] Apparently the Iranians were apprehensive about this kind of stalemate. They felt that with the Red Army in occupation of their north, the process of procrastination would win the secessionist movement a *de facto* recognition. They therefore decided to buy their national integrity with an oil concession.

On April 4, 1946, the Prime Minister of Iran and the Soviet Ambassador in Tehran issued a joint communiqué in which they declared an agreement had been reached between the two governments. The agreement (a) called for the date of May 6, 1946, to be the last of Russian troops withdrawal; (b) recognized

Azerbaijan as an internal Iranian affair; and (c) organized a Russo-Iranian Oil company for a period of fifty years. The agreement further stipulated the ratification of the oil concession by the Majlis should be within seven months from March 24.[40]

On April 14, 1946, the Iranian delegate to the United Nations was instructed to drop the case, and it was later reported that the Russians had their last soldier out of Iran by May 9, 1946.[41]

The Soviets had demanded the ratification of the agreement seven months from March 24. At the time of signing the agreement, however, the 14th Iranian Majlis had already been dissolved on March 11, 1946. Before its dissolution, the 14th Majlis saw fit to pass a bill to prevent the government from holding the election while foreign troops were on Iranian soil. The Soviets were out by May 1946, yet the Iranian government showed no intention of calling for the election of the 15th Majlis. When it did, it was already January, 1947, and when the Majlis was finally convened by the Shah, it was July 17, 1947 – seventeen months after March 14, 1946.[42]

On the night of September 11, 1947, three days before the Russian concessions were to be put before the Majlis, and while the members of the Majlis had been puzzling over the United States stand, United States Ambassador to Tehran, George V. Allan, speaking before the Iranian-American Relations Society, 'cleared the picture with one of his diplomatic brushes.' The Ambassador declared:

> The United States is firm in its conviction that any proposals made by one sovereign government to another should not be accompanied by threats or intimidation.

The Ambassador went on to say:

> Patriotic Iranians, when considering matters affecting their national interests, may therefore rest assured that the American people will support fully their freedom to make their choice.

He then concluded his speech in this manner:

> Iran resources belong to Iran. Iran can give them away free of charge or refuse to dispose of them at any price if it so desires.[43]

This definition of United States policy was long awaited in Iran and was 'expected to have a widespread effect.'[44] Naturally, the Soviets were enraged at the direction of Iranian policy, and they protested the action vigorously. The British, too, were disturbed, but it was believed their disturbance emanated from their being unhappy over Soviet frustration. This contention, however, was not borne out by the events of the days to come, for by a vote of 102 to 2 the Majlis, on October 22, 1947, rejected the ratification of the Russian concession. They also inserted a final paragraph in that rejection:

> In all cases where the rights of the Iranian nation, in respect of the country's natural resources, whether underground or otherwise, have been impaired, particularly in regard to the *southern* oil, the government is required to enter into such negotiations and take such measures as necessary to regain the national rights and inform the Majlis of the result.[45] (Emphasis added.)

From now on the problem is not of a Soviet concession in the north, but rather of a huge British oil establishment in the south. The above law was ratified by the Shah on November 5, 1947, and the first step in a long series of power maneuvers, agitation, and nationalization was taken.

Empowered by the new law, the Iranian government opened negotiations with representatives of the Anglo-Iranian Oil Company. It was by then obvious that something should be done to close the gap between the company's payments to the British government and its payments to the Iranians. Almost everyone in an official capacity – American, British, or Iranian – was pronouncing discomfort at the way the company was distributing its profits. According to Dr Henry F. Grady, United States

Ambassador to Tehran during the crisis, the company made £100,000,000 in net profits in 1950 and paid Iran £13,000,000 to £15,000,000 a year in royalties – about 15 percent.[46] Earlier, Sir Anthony Eden, then in the Opposition, had commented on the piling up of retained earnings rather than distributing the profits in dividends so as to enable Iran to get something.

The Iranians, too, had their opinion on the financial grievances. Though they were barred from having access to the company's records, it was not difficult for them to arrive at an accurate estimation. By using international market prices and knowing the amount of tonnage produced, the representative of Iran to the United Nations could declare that in 1950 alone the 'Company derived a profit of between $500 million and $550 million, or between £180 million and £200 million, from its oil enterprises in Iran.' 'Of this huge amount,' the Iranian representative observed, 'Iran received only $45 million, or about £16 million as royalties, share of profits, and taxes.'[47] In answering a comment from the representative of the United Kingdom to the United Nations, the Iranian representative declared that 'the profits of the Company in the year 1950 alone, after deducting the share paid to Iran, amounted to more than the entire sum of £114 million cited by the representative of the United Kingdom as the total sum paid to Iran in royalties in the course of the past half century.[48]

By now the Iranians were aware of the 50-50 formula (equal sharing of profits and losses) which was secured by the Venezuelan Government.[49] They hoped, therefore, that their negotiations with the British would produce similar results. The Iranians were also hoping that the result of their negotiations would subject the company's imports and exports to Iranian custom duties, eliminate 'the secret special price charged by the Company for oil sold to the British Navy,' and calculate Iranian receipts before payments of taxes to the British government. They also expected to see an increase in the number of Iranians employed by the company, the employment of Iranians in managerial posts, and the exertion of some control over the accounting practices of the company.[50]

These then were the Iranian expectations. Two years of negotiations, however, produced an agreement, referred to as the supplement agreement, which was, at best, not enough. All told, the supplement agreement, signed July 17, 1949, amounted to an increase in the government's share of net profits (after British taxes), from about 15-20 percent to 25-30 percent.[51] According to Dr Laurence Lockhart, even if this agreement 'had come into force, there would still have been an appreciable gap between the receipts of the Persian government and those of the British Treasury.'[52]

Nevertheless, the supplement agreement was signed by the government and presented for ratification to the Majlis which was due to dissolve in ten days, July 28, 1949. No action was taken by the Majlis, and the agreement had to await ratification by the 16th Majlis which was duly assembled in the middle of June 1950. The election of the 16th Majlis brought back most of the familiar faces, including Dr Mohammad Musaddeq and his colleagues.[53]

Ali Mansur was named Prime Minister of Iran in March 1950. By then (February 20, 1949) it was known that Saudi Arabia had concluded an agreement with Pacific Western Oil Corporation, a newcomer, which would give Saudi Arabia twice as much as the supplement agreement would give Iran.[54] The Prime Minister feared precipitation of a crisis in the Majlis and tried to renegotiate the agreement with the company in the light of the 50-50 formula. The company, however, 'stood firm, on legal ground, that until the Majlis had rejected the supplementary agreement no change could be made.'[55] The attitude of the company forced the resignation of the Prime Minister.

On June 26, 1950, General Ali Razmara, the popular Chief of Staff, became Prime Minister of Iran. He, too, tried to obtain additional concessions of non-monetary character, but 'the British attitude, during the summer and fall of 1950, was seriously obtuse and complacent.'[56] The British then persuaded the Prime Minister to present the agreement to the Majlis. It was promptly rejected, and the rejection forced the resignation of the Finance Minister. Subsequently, the agreement was referred to a special committee headed by Dr Musaddeq which, in December 1950, rejected it on the grounds that it did not safeguard the interest of the country.

The committee was then charged with the drawing up of guidelines for the government to follow.[57]

It is worth mentioning at this point that in their attempts to have the supplement agreement renegotiated, the Iranian government had the open support of Dr Henry Grady, United States Ambassador to Tehran.[58]

So far the idea of nationalization was not seriously entertained, though it was reported that in October, 1950, a deputy in the Majlis put forward the idea for consideration.[59] Prior to this, however, the idea had been signaled by Dr Grady when in mid-June, 1950, he called upon the local manager of Anglo-Iranian and advised him to get in touch with his superiors in London to send one of their top officials to deal with the Iranians before it was too late. Grady drew 'the local manager's attention to the dangers of a move toward nationalization, based upon the experience of American oil companies in Mexico in 1938.' Grady volunteered this advice 'even before Mussadeq's oil commission in the Majlis began to discuss the question.'[60]

Other than this signal, there was, according to R.H.S. Crossman, member of the British Parliament, 'a great deal of whispering by Americans that if the British were got rid of the Persians would find available to them American technicians.' Crossman went on to say: 'As we know, the whisper is everything in the Middle East. What is significant is that the Persians were led to believe that if they did not sign the agreement with Anglo-Iranian they could get a better deal from one of the American oil companies.'[61]

The big step towards the cause of nationalization was launched however, when, in December 1950, it became known that the Arabian American Oil Company (Aramco) had concluded with neighboring Saudi Arabia a 50-50 agreement whereby Saudi Arabia would get 50 percent of the profits of the company as taxes and royalties.

By then the British were informed and they communicated to the Iranian government their readiness to negotiate a similar agreement in line with the one concluded by Saudi Arabia. The offer was too late. The special committee on oil under the leadership of Musaddeq had already, on February 19, 1951, proposed to the Majlis the nationalization of the oil industry. Premier Razmara then consulted his Iranian experts who doubted the practicability of nationalization. Accordingly, on March 3, 1951, General Razmara rejected the proposal. On March 7, he was shot dead.

The assassination of General Razmara silenced the voice of moderation. The movement for nationalization was 'fast gaining momentum.'[62] On March 8, the special committee adopted the nationalization proposal, and on March 15 the Majlis approved it. Hussein Ala, a former Iranian Ambassador to Washington, succeeded General Razmara. By then, however, tension was mounting, and the country was experiencing a general strike. The government proclaimed martial law. However, on April 27 Hussein Ala was forced to resign, and on April 28 the Majlis acclaimed Mohammad Musaddeq premier.[63] On April 30, 1951, a nine-article Enabling Law for the implementation of nationalization was passed by the Majlis.[64]

The British protested the act of nationalization and sent to Iran a delegation under the chairmanship of Basil Jackson, deputy chairman of the Board of Anglo-Iranian. The British made an offer which was rejected by the Iranians. On May 26 the British filed a petition with the International Court of Justice. The Iranian government, considering the dispute to be one of domestic affairs, declined to recognize the Court's jurisdiction. The situation was getting worse every day, and by September, 1951, the country was in a state of chaos. The oil operation was brought to a standstill, British personnel were ordered to leave the country, and Iran severed its diplomatic relations with Britain. On October 1, 1951, the British brought the dispute to the United Nations Security Council, which decided to withhold its decision pending the International Court pronouncement on its competence.

The United States was in complete support of the nationalization cause. At the peak of the crisis the London *Times* stated that an American spokesman from the State Department 'drew the attention to the fact that Standard Oil Company of New York and Standard Oil Company of New Jersey had an agreement to buy 40 percent of the Anglo-Iranian Company's Oil.'[65] Such an announcement at such a time served to get the message to the Iranians that if the British were to boycott the Iranian oil, American companies would stand ready to help in alleviating the economic hardship by mitigating the British blockade. In fact, the *Times* did not leave to its readers much speculation when it put the preceding statement into perspective and immediately qualified it by saying that 'Mr McGhee, the United States Assistant Secretary of State, is on his way to Teheran from Karachi to confer an the situation tomorrow with the American Ambassador.'[66]

George C. McGhee, Assistant Secretary of State for Near Eastern Affairs, himself an oilman,[67] arrived at Tehran on March 17, 1951. While there, he was known to have criticized the behavior of Anglo-Iranian Oil Company. In answering the charge, Crossman, introducing McGhee in the parliamentary debates as an 'Oklahoma oil tycoon,' stated that whether McGhee 'was right or wrong in his criticism of Anglo-Iranian, the impression he made on the Persians was that if the British were kicked out they could rely on somebody else and they might do a little better.'[68] George Wigg, another member of Parliament, carried the notion further when he stated that McGhee 'gave the Persians the impression that, as far as American policy was concerned, they could go ahead with nationalization.'[69]

The London *Times* followed the debates in the House of Commons and reported that the British Secretary of State for Foreign Affairs, Herbert Morrison, 'was at pains to clear the air over this matter.' The *Times* cited the Secretary as saying that 'the United States Government was in general support of the line we had taken... However,' continued Morrison, 'there had been some people, not of outstanding importance, associated with the American oil industry, who had said foolish and perhaps dangerous things.'[70]

Upon his return from Iran, and on the eve of his talks with the British delegation on the Iranian crisis, George McGhee told a CBS radio audience that he 'had the opportunity for a long discussion with the Shah in Teheran.' He continued:

> I conveyed to the Shah the confidence which our
> Government has in him, and the fact that we are fully behind
> Iran and want to do what we can to assist Iran. ... The point I
> want to make is that you do not succeed in any endeavor by
> exaggerating the difficulties that lie ahead. You capitalize on
> what you have and drive ahead with a will to win.[71]

The question before the Assistant Secretary, of course, was to win
against whom? For he himself had said that 'there is no indication
that the Kremlin engineered the present crisis in Iran.'[72]

Indeed, this last statement by McGhee was immediately
picked up by the London *Times* and presented in a manner which
raised questions. Now, implied the *Times*, if there is no indication
that the Kremlin engineered the crisis, who would have engineered
it? In a direct reference to the 50-50 formula concluded between
Saudi Arabia and the American companies, the *Times* chose to
clarify the puzzle in this manner:

> There has been a tendency in America, since the
> nationalization legislation was passed by the Majlis, to read
> the British government and Anglo-Iranian Oil Company
> tendentious little lectures in which the excellent relations
> between Arabian-American Oil Company and the authorities
> in Saudi Arabia are held up as an example. In the past few
> days it has been reported that the State Department had been
> telling the Foreign Office that it was convinced that the
> decision to nationalize was irrevocable and that protests
> against it were useless.[73]

On April 9, McGhee had a meeting with the Iranian
Ambassador to Washington, Nasrollah Entezam. After the meeting
the Ambassador told newsmen that there was little possibility that
his government would reverse its decision to nationalize
petroleum, including the Anglo-Iranian Oil Company.[74]

Later in the day, high officials of the British Foreign Office,
headed by the British Ambassador to Washington, Sir Oliver
Franks, and representatives of the Anglo-Iranian Oil Company

conferred with McGhee and other officials of the US Department of State. A three-hour meeting produced a communiqué, which reads in part:

> The two Governments, while recognizing that the question relating to Iranian oil must be settled elsewhere, have deemed it advisable to exchange view informally. Further exchanges of views will take place.[75]

It is obvious from the wording of the communiqué that no agreement had been reached. Then Gulf Oil Corporation's Charles Hamilton said:

> The British delegates came prepared to give Iran a larger share in Angle-Iranian Oil Co.'s profits and to admit some Iranians to their board. However, they were chagrined when they learned the American attitude envisaged that any solution to the Iranian oil problem must be within the framework of Iran's announced desire to nationalize its oil industry. The British refused to accept the *principle* of nationalization and so went home.[76] (Emphasis added.)

No one, it appeared, except perhaps officials of the US Department of State, really knew the exact meaning of 'in principle' – nationalization in principle. The British rejected the idea and went home. For them, nationalization meant ownership, management, control, and distribution of the oil by the Iranians. Nationalization meant exactly what George V. Allan, United States Ambassador to Iran, had told the Iranians four years before: 'Iran's resources belong to Iran. Iran can give them away free of charge or refuse to dispose of them at any price if it so desires.'[77]

Gradually, however, the British came to grips with the meaning of the kind of nationalization 'in principle' the United States envisaged. They knew, for example, that it amounted to bestowing an overall title of ownership upon the Iranians, but to keep the management, the control of production, the actual operation, and the distribution of the oil in the hands of the company until the end of the contract.

The British, of course, saw no basic change between their previous organizational set-up and the one set out by the United States. Yet they were concerned about the interruption of the oil flow from Iran. When, however, they learned of an agreement, dated June 25, 1951, concluded among the American companies 'to set up machinery to close the gap, created by the loss of Iranian oil,'[78] they sighed with relief! The British felt that they were no longer under the pressure of an oil cut-off and, hence, decided 'to fight it out with the Iranians to the very end.'[79] Accordingly, they took all the time available in submitting, severally and jointly with the United States, one proposal after another, looking for a solution to the problem, all of which did not go beyond the meaning of nationalization 'in principle' and were therefore rejected by Iran.

The Iranians on their part thought they had it made and that their half-a-century of hard economic exploitation was almost over. They too, however, got around the real meaning of nationalization 'in principle.' By then, however, it was too late. 'The oil matter,' said Dr Grady, 'had become a symbol of the passion for complete economic independence.'[80]

'Nationalize the oil industry!' had been picked up by thousands throughout the country. Supporting Prime Minister Musaddeq were the various religious organizations, the nationalists, and the communists-each for his own reason. The cause of nationalization had not only the overwhelming support of the Iranians, but also the unqualified support of the United States – a very important leverage in power politics. Besides, Dr Musaddeq was a man of character. According to Dr Laurence Lockhart, a distinguished historian who worked for the Foreign Office and the Anglo-Iranian Oil Company and who carried out research work on Persian history, 'Musaddeq, is a man of sincerity and genuine patriotism.'[81] Lockhart attested that not only Musaddeq's astuteness and obstinacy were underestimated but also his pertinency and consistency. Musaddeq, Henry Grady wrote:

is a man of great ability as a popular leader and is regarded even by his critics as thoroughly honest. He is also a man of great intelligence, wit and education – a cultured Persian gentleman. He reminds me of the late Mahatma Gandhi. He is a little old man in a frail body, but with a will of iron and a passion for what he regards as the best interests of his people.[82]

Now, with this overwhelming support from inside and outside, and with this man of principles behind the cause of nationalization, a retreat from the real meaning of nationalization was political suicide. Dr Musaddeq thus became a prisoner of his principles and patriotism, and he had to turn down every offer that fell short of complete economic independence.

Musaddeq rejected the proposals of Basil Jackson, deputy chairman of the board of Anglo-Iranian. He then rejected the proposals of Richard Stokes, Lord Privy Seal, made in cooperation with W. Averell Harriman, President Truman's special envoy. Finally, he rejected the proposals submitted under the names of President Truman and Sir Winston Churchill.

These proposals were rejected because they all were revolving around the 'principle' of nationalization. They resembled and were governed by the structural organization of the idea of the Consortium. The idea of the Consortium, according to Howard W. Page of Standard Oil of New Jersey and head of the negotiating team that went to Iran after the fall of Musaddeq, 'is the same as if someone gave you full use of his car for a specific period, instead of selling it to you for the same period. True, you don't own it. But, in reality, you get much the same results. And you can get the finest lawyers to argue themselves silly over the legal distinctions'.[83]

There are those who believe that Musaddeq's rejection of so many offers was founded on his capitalization on America's fear that Iran would slip into the Soviet sphere of influence.[84] Although there is some truth in this belief, the facts of history do not bear it out. Throughout the ages, Iran had been known for her continuous resistance to the Russians – Czarist, Communist or otherwise. We know, for example, that with the help of the United States the

Iranian Majlis scorned the Russian contract, and Premier Qavam dismissed his three Communist ministers without the Soviet Union being able to retaliate. Olaf Caroe would say that 'a Muslim State may be less likely to succumb to Communist penetration in the absence of the Red Army than many States in Europe.'[85]

Ideologically, the Soviet Union has an even harder time to convert a Muslim state into Communism. According to Halford Hoskins, the people of the Middle East have a 'natural distaste' for Communist ideology.[86] Walter Laqueur comments upon the evolution of Communism in the Middle East in this manner:

> The Communist parties have no chance to become mass parties unless they adjust much more to specific local conditions. This means (to give but a few examples) that the communists would have to pay more than lip service to Islam and Arab nationalism, that they disavow dialectical materialism and that they give up any idea of nationalizing agriculture. It means, in other words, that the communists cease to be communists and transfer themselves into nationalist-socialists.[87]

This state of mind of the Middle Eastern people was certainly known to McGhee when he told his CBS radio audience that 'without exception the vast majority of the people in the Near East and South Asia abhor the Communist doctrine.'[88]

This being the case and noting that the fervor for nationalization was going beyond what it set out to accomplish, the United States came to the aid of the British in their economic blockade to force Musaddeq out:

> The Americans came around to the British point of view that the deterioration of economic and financial conditions in Iran would force Musaddeq out, rather than bringing the Soviets in, and that a new government would be amenable to a sensible solution.[89]

The first instance in which the United States came into the open in direct support of the British economic policy was on June 27, 1951, when US Secretary of State Dean Acheson, strongly criticized Iran's policy on the oil dispute as one of threat and fear:

This atmosphere of threat and fear which results from hasty efforts to force cooperation in the implementation of the nationalization law cannot but seriously affect the morale of employees and, consequently, their willingness to remain in Iran.[90]

Prior to Acheson's statement, representatives of the major international companies were called on April 25, 1951, by the Petroleum Administration for Defense to 'discuss what could be done to meet world oil demands if Iran closed down.'[91] The result of this meeting was the formation of the agreement of June 25, 1951, among the American oil companies. The conferees, however, were worried that their agreement was a cartel-type arrangement which would fall under the jurisdiction of Antitrust Law. Therefore, 'pressure was applied in the name of national defense to get the United States Department of Justice to promise that there would be no persecution for antitrust violation.'[92]

Accordingly, when Iranian production was stopped, the loss of 650,000 barrels of oil per day was made up for by production from Texas and Saudi Arabia. The major oil companies then agreed not to 'touch a barrel of Iranian-produced oil.' Some of the independent American oil companies wanted to exploit the Iranian situation, 'but on the whole the American government consistently discouraged such attempts.'[93]

By the autumn of 1951, Iran was in economic despair. Dr Grady was rather puzzled over the American cooperation with the British. He said that 'while our hesitation about granting the long overdue credit might have had nothing to do with the oil situation, there was sufficient parallelism and timing between British and American action to make the Iranians feel certain we were following the British in putting on financial squeeze.' He went on to conclude that 'one of the most important problems now facing us is the thorough-going coordination of British and American policy.'[94]

While in the United States pleading the Iranian case before the United Nations Security Council, Dr Musaddeq met, on October 23, 1951, with President Truman and appealed for a loan of $120 million to help Iran break the British economic siege. The appeal was of no avail, however.[95]

Two years of economic and financial strangulation brought Iran to a state of turmoil. Apparently desperate, Prime Minister Musaddeq wrote President Eisenhower on January 9, 1953, eleven days before Eisenhower's inauguration, and again on May 28, 1953. Musaddeq hinted at the British-American collaboration against Iran and then went on to say that if the United States could not apply its pressure on the British, then let the United States extend its 'effective economic assistance to enable Iran to utilize her other resources.'[96] On June 29, 1953, President Eisenhower wrote Dr Musaddeq noting that 'it would not be fair to the American taxpayers for the United States Government to extend any considerable amount of economic aid to Iran so long as Iran could have access to funds derived from the sale of its oil and oil products if reasonable agreement were reached.'[97]

As the economic situation continued to worsen, Musaddeq's position was getting rather precarious. By August 1953, tensions were mounting in Iran, and demonstrations for and against Musaddeq were breaking out all over the country. Driven to the extreme, he forced the Shah to flee the country. On August 19, 1953, the Shah's forces gathered momentum and arrested the Prime Minister. General Fazlollah Zahedi assumed the premiership and asked for American help, which was forthcoming in the form of $45,000,000 emergency grant-in-aid in addition to $23,400,000 under the technical assistance program.[98]

The British finally had it made. They rejoiced over the fall of Musaddeq. They therefore were patiently waiting for the Iranians to appeal to them when, suddenly, Herbert Hoover, Jr[99] appeared in London in September, 1953. His appearance 'was met in many British quarters by resounding silence.'[100]

While in Britain, Hoover presented the British with the idea of an international consortium. The idea of the consortium provided that 'the oil properties would have to stay nationalized, and the British could not return to the oil fields in a clearly *dominant* position.' (Emphasis added.) The idea was further expanded upon so as to have Iran 'transfer full operating rights to the properties to some new, politically acceptable international oil group for a long-term period roughly corresponding to the old concession.'[101]

According to Charles Hamilton, the American government envisioned the limitation of the British share in the consortium to be 40 percent. A similar share would go to the Americans, with the remaining 20 percent to be prorated between the Dutch and the French (see Table 13).

What about Iran? Or, to put the question differently, where did she fit into the distribution of the spoils after these years of turmoil? Iran was given a 'nationalization in principle.' She was asked to sign a contract to grant exclusive use and complete management of her properties to the consortium for 40 years. According to a puzzled Tehran newspaper cited by Wanda Jablonski: 'The consortium will act as Iran's agent, only Iran won't be able to tell the agent what to do for 40 years.'[102]

Clearly the British were unhappy with these arrangements. They saw their last stronghold invaded by the Americans, and their 100 percent exclusively British trust in Iran was turned into a minority holding. It is not known what kind of pressure had been applied to the British to yield to the idea of the consortium. Hamilton, however, offers in explanation that 'largely because of American persistence at the governmental level, the Anglo-Iranian wreck was salvaged to live again as British Petroleum.'[103] One would think that the Americans threatened the British with immediate help to Iran – directly and indirectly; directly, through the extension of immediate loans and aids; and indirectly, through allowing American companies to begin the purchasing and marketing of Iranian oil.

TABLE 13
The Consortium

	Percentage Share	
British interest		40%
British Petroleum Co. (formerly Anglo-Iranian Oil Co.)		
American interest		40%
Standard of New Jersey	7%	
Standard of California	7%	
Gulf Oil Corporation	7%	
Socony-Mobil	7%	
Texaco Inc.	7%	
American Independent Oil Co.	5%	
British-Dutch interest		14%
Royal Dutch-Shell qroup		
French interest		6%
Compagnie Francaise des Petroles		

Source: 'Basic Analysis, Oil,' Standard and Poor's Industry Surveys, *Section 2 (December 11, 1969), O 55.*

In the process an agreement was reached between the Iranian government and members of the consortium. The agreement was signed on August 5, 1954, and ratified by the Majlis in October. The agreement that ended the dispute amounted to a new concession contract. It founded two companies under the Dutch laws for operating the oil industry in Iran. The agreement was to run for twenty-five years, with the right of renewal for three five-year extensions at the option of the consortium under specific conditions.[104]

The agreement provided Iran with the already prevailing pattern of equal distribution of profits and royalties among the oil-producing countries and the companies. Also, the agreement recognized the National Iranian Oil Company (set up in 1951 by Iran to take over Anglo-Iranian), but it limited its operation to marketing outlets in Iran only. In addition, the National Iranian Oil Company was assigned the provision of services of a non-industrial nature.[105]

The agreement with Iran was in two parts. The first governed future relations between the consortium and Iran, and the second part specified the terms for the settlement of disputes. In this second part Iran was to pay Anglo-Iranian £66 million claimed by the Anglo-Iranian for losses incurred due to the nationalization of the oil industry. In addition, Anglo-Iranian was to be paid by the American companies, Shell, and Compagnie Francaise the awesome amount of $600,000,000, apparently for the right to share in the riches of Iran. One, however, should not exaggerate the amount and the difficulty Iran would have had in finding the money had she succeeded with nationalization. For the seven members of the consortium paid Anglo-Iranian $90,000,000 in three equal installments over a period of twelve months. The rest, $510,000,000, was to be paid to Anglo-Iranian by allowing it ten cents on every barrel of crude oil and products exported from Iran by the seven companies until the amount had been fully paid.[106]

It now appears, in historical perspective, and after seeing the terms of the settlement, that the whole Iranian fiasco was just another manifestation of the struggle among the world's major powers over the control of oil.

Endnotes

[1] Edward Ashley Bayne, 'Crisis of Confidence in Iran,' *Foreign Affairs*, XXIX, No. 4 (July, 1951), 580.

[2] Zuhayr Mikdashi, *A Financial Analysis of Middle Eastern Oil Concessions 1901-65* (New York: Frederick A. Praeger, 1966), 74.

[3] Charles W. Hamilton, *Americans and Oil in the Middle East* (Houston, Texas: Gulf Publishing, 1962), 34 and 36.

[4] League of Nations, *Official Journal*, XIV[1] (February, 1933), 300-303.

[5] Ibid., 291.

[6] Ibid., 302.

[7] Hamilton, op. cit., 38; Mikdashi, op. cit., 76.

[8] Mikdashi, op. cit., 77.

[9] Great Britain, Parliament, *Parliamentary Debates, House of Commons*, Fifth Series, CCCCLXXXIX, (June 21, 1951), col. 752.

[10] Ibid.

[11] Hamilton, op. cit., 38.

[12] Benjamin Shwadran, *The Middle East, Oil and the Great Powers* (New York: Frederick A. Praeger, 1955), 55-56.

[13] Hamilton, op. cit., 38.

[14] Laurence Lockhart, 'The Causes of the Anglo-Persian Oil Dispute,' *Journal of Royal Central Asian Society*, XL (April, 1953), 144-145.

[15] Great Britain, Parliament, *Parliamentary Debates, House of Commons*, Fifth Series, CCCLXXIV (September 30, 1941), col. 518.

[16] United Nations, Security Council, *Official Records*, 1st Year, 1st Series, Supplement No. 1 (January-February, 1946), 43-48.

[17] Ibid.

[18] Ibid.

[19] Ibid., 49-50.

[20] Quoted in T.H. Vail Motter, *US Army in World War II, The Middle East Theater: The Persian Corridor and Aid to Russia*, Office of the Chief of Military History, Department of the Army (Washington DC: Government Printing Office, 1952), 472.

[21] J.C. Hurewitz, *Middle East Dilemmas: The Background of United States Policy* (New York: Harper and Brothers, 1953), 23.

[22] Ibid., 24.

[23] Motter, op. cit., 290.

[24] Ibid., 285 and 289.

[25] Ibid., 288-289.

[26] Quoted in ibid., 289.

[27] George Edward Kirk, *The Middle East in the War, Survey of International Affairs 1939-1946* (London: Oxford University Press, 1952), 272.

[28] Cordell Hull, *The Memoirs of Cordell Hull*, Vol. II (New York: The Macmillan Company, 1948), 1508-1509.

[29] Ibid., 1509; Kirk, op. cit., 474.

[30] Hamilton, op. cit., 40-41; Kirk, op. cit., 474.

[31] Hamilton, op. cit., 40.

[32] Quoted in Kirk, op. cit., 475.

[33] Hull, op. cit., 1509.

[34] Lockhart, op. cit., 143.

[35] Quoted in Kirk, op. cit., 476.

[36] Herbert Feis, 'Oil For Peace Or War,' *Foreign Affairs*, XXXII. No. 3 (April, 1954), 417.

[37] United Nations, Security Council, *Official Records*, 1st Year, 1st Series, No. 2, (March-June, 1946), 64

[38] Ibid., 64-65.

[39] Ibid., 74 and 88-89.

[40] Arthur Chester Millspaugh, *Americans in Persia* (Washington DC: The Brookings Institute, 1946), 196-197; Shwadran, op. cit., 78-79.

[41] United Nations, Security Council, *Official Records*, 1st Year, 1st Series, No. 2 (March-June, 1946), 64.

[42] Hurewitz, op. cit., 28-29.

[43] 'US Bids Iran Resist Threats as Debate on Soviet Oil Nears,' *The New York Times* (September 12, 1947), 1 and 8.

[44] Ibid., 1.

[45] Quoted in Shwadran, op. cit., 13.

[46] Henry F. Grady, 'What Went Wrong in Iran?' *Saturday Evening Post*, CCXXIV (January 5, 1952), 57.

[47] United Nations, Security Council, *Official Records*, Sixth Year, 563rd Meeting (October 17, 1951), 15.

[48] Ibid.; Lockhart, op. cit., 145.

[49] Bayne, op. cit., 581.

[50] Grady, op. cit., 57: Bayne, op. cit., 581; Shwadran, op. cit., 147.

[51] Hurewitz, op. cit., 35; Bayne, op. cit., 581; Grady, op. cit., 57.

[52] Lockhart, op. cit., 145.

[53] Bayne, op. cit., 582.

[54] United Nations, Security Council, *Official Records*, Sixth Year, 563rd Meeting (October 17, 1951), 19-20; Mikdashi, op. cit., 136.

[55] Bayne, op. cit., 583.

[56] Grady, op. cit., 57-58.

[57] Shwadran, op. cit., 105.

[58] See Grady, op. cit., 109; Lockhart, op. cit., 147.

[59] Lockhart, op. cit., 146

[60] Grady, op. cit., 58.

[61] Great Britain, Parliament, *Parliamentary Debates, House of Commons*, Fifth Series, Vol. 489 (June 21, 1951), col. 776.

[62] Lockhart, op. cit., 149.

[63] Bayne, op. cit., 580.

[64] Shwadran, op. cit., 108.

[65] *The Times* (London), March 17, 1951, 4.

[66] Ibid.

[67] Shwadran, op. cit., 148.

[68] Great Britain, Parliament, *Parliamentary Debates*, House of Commons, Fifth Series, Vol. 489 (June 21, 1951), cols. 776-777.

[69] Ibid., col. 801.

[70] *The Times* (London), (June 22, 1951), 6.

[71] United States, Department of State, *Bulletin*, Vol. 24, No. 616 (April 3, 1951), 657.

[72] Ibid.

[73] 'US Policy on Persian Oil,' *The Times* (London), (April 10, 1951), 3.

[74] 'Iran Accord Asked by US and Britain,' *The New York Times* (April 10, 1951), 13.

[75] United States, Department of State, *Bulletin*, Vol. 24, No. 616 (April 23, 1951), 661.

[76] Hamilton, op. cit., 48.

[77] 'US Bids Iran Resist Threats as Debate of Soviet Oil Nears,' *The New York Times* (September 12, 1947).

[78] US, Congress, Senate and House, Select Committee on Small Business of the Senate and the House of Representatives, *A Report: The Third Petroleum Congress*, 88th Cong., 2nd Sess. (Washington DC: Government Printing Office, 1952), 11.

[79] Shwadran, op. cit., 149.

[80] Grady, op. cit.

[81] Lockhart, op. cit., 141.

[82] Grady, op. cit., 58.

[83] Quoted in Wanda Jablonski, 'Master Stroke in Iran,' *Collier's* (January 21, 1955), 34 and 36.

[84] George Lenczowski, *The Middle East in World Affairs* (3rd edn.; Ithaca, New York: Cornell University Press, 1962), 212; Shwadran, op. cit., 150-151.

[85] Olaf Caroe, *Wells of Power, the Oilfields of South-Western Asia, A Regional and Global Study* (London: Macmillan and Co., Ltd., 1951), 75.

[86] Halford L. Hoskins, 'Needed: A Strategy for Oil,' *Foreign Affairs*, XXIX, No. 2 (January, 1951), 236.

[87] Walter Laqueur, 'Russia Enters the Middle East,' *Foreign Affairs*, XXXXVII, No. 2 (January, 1969), 298.

[88] United States, Department of State, *Bulletin*, XXIV, No. 616 (April 23, 1951), 657.

[89] Shwadran, op. cit., 152.

[90] United States, Department of State, *Bulletin*, XXV, No. 628 (July 9, 1951), 73; 'US Asks Persia to Modify oil Policy,' *The Times* (London) (June 28, 1951), 6.

[91] US, Congress, Senate and House, Select Committee on Small Business of the Senate and the House of Representatives, *A Report: The Third Petroleum Congress*, 88th Cong., 2nd Sess. (Washington DC: Government Printing office, 1952), 12.

[92] Ibid.

[93] Ibid., 11; Shwadran, op. cit., 150.

[94] Grady, op. cit., 57-58.

[95] Hamilton, op. cit., 54; Shwadran, op. cit., 131-132.

[96] United States, Department of State, *Bulletin*, XXIX, No. 773 (July 13, 1953), 75.

[97] Ibid.

[98] Lenczowski, op. cit., 203.

[99] Hoover was an American oil consultant who worked for the Iranian Government as a financial advisor during World War II. He was chosen by the Secretary of State, John Foster Dulles, for a fact-finding mission in the oil dispute. Later, in 1955, he was appointed United States Under-Secretary of State.

[100] Jablonski, op. cit., 34.

[101] Ibid.

[102] Quoted in ibid.

[103] Hamilton, op. cit., 59.

[104] *The Middle East and North Africa, 1969-70* (16th edn.; London: Europa Publications Limited, 1969), 274.

[105] Christopher Tugendhat, *Oil: The Biggest Business* (New York: G.P. Putnam's Sons, 1968), 143-144; Mikdashi, op. cit., 147.

[106] Shwadran, op. cit., 188; Mikdashi, op. cit., 159.

5

Instruments of Control

The Built-in Dispute

Britain controls about one-third of the total oil production of the Persian Gulf area and draws about 60 percent of her crude oil requirements from that region. She draws most of her petroleum requirements from 'sterling' oil and, accordingly, the relief of her economy from the heavy drain on foreign exchange is obvious.[1]

The Center for Strategic and International Studies put Britain's annual earnings from her oil operations in the Gulf at about $500 million.[2] Total British stake – investments, return thereon, and revenues from sales – in the region was in 1970 estimated at £1,200 million.[3]

With this economic involvement in mind, it was then natural for Robert G. Landen to note that 'the continued economic viability of Britain itself was sometimes pictured as being intimately tied to the Persian Gulf oil operations.' Along the same line, David H. Finnie observed that without the Gulf oil resources, 'Great Britain, and indeed Western Europe might well be unable to survive in its present form.'[4]

Despite this economic stake, Britain's Labour government found it safe to announce on January 16, 1968, its intention to withdraw all British forces 'East of Suez,' including those in the Persian Gulf, by the end of 1971. British military forces in the Persian Gulf were then estimated at 6,000 soldiers stationed on the island of Bahrain, supported by an air base at Sharjah on the Omani Coast and eight ships of the Royal Navy.[5]

Immediately after its announcement, the decision to withdraw was the subject of severe criticism from almost every direction.[6] Although out of office, the British Conservative Party attacked the Labour decision and promised retention of British forces in the Persian Gulf if they should win the election.[7] The June 1970, election did bring the Conservatives into power. Sir William Luce, former Governor and Commander-in-Chief of Aden and former Political Resident, Persian Gulf, was sent to the countries of the Gulf for an eight-week fact-finding mission. Reluctantly, afterward, the Conservatives announced their adherence to the Labour plan of ending the British military presence in the Gulf by the end of 1971.[8]

The major criticism leveled against the British withdrawal centered around the emergence of a potential power vacuum which would be bound to create a persistent state of internal unrest and disorder. The critics see that conflicts between Arabs and Arabs and between Arabs and Iranians, escalated and exploited by hostile powers, would necessarily lead to interruption of the oil flow.[9] The criticism is not, of course, without its justification. What is ironical, however, is that through conflicts and fragmentation an added insurance against oil interruption is reasonably provided. It stems from an old British policy of 'divide and rule' or to use the words of D.C. Watt, Senior Lecturer in International History at the London School of Economics, the policy of 'balanced tensions and conflicting claims.'[10]

Before we look more closely at this concept and the British role in widening the fragmentation and intensifying the conflict, let us first consider another major problem said to be contingent upon the British withdrawal.

It has been said, for example, that a British withdrawal would necessarily pave the way for direct Soviet intervention. The British, however, had reasoned that the 'main Soviet purpose is not to attack the free world's shipping but to establish political influence in countries bordering the area.'[11] As in Iraq, where the Russians are more influential, the largest portion of Iraqi oil flows West. If Iraq chooses other than this direction for her oil, Iran or Kuwait or Saudi Arabia or Qatar or Bahrain or Abu Dhabi or Dubai or Oman would be able to fill the gap. More importantly, though, the 6,000 British soldiers stationed on the island of Bahrain could not prevent the Soviet Union from gaining influence in Iraq. Nor was the British military presence relevant to the conclusion of an agreement between Iran and the Soviet Union under which the development of the huge resources of Iranian natural gas became possible.

Soviet influence may even be considered as a plus factor in a policy of widening fragmentation as added insurance against oil interruption. Soviet influence has the advantage of controlling the conflict for fear of giant involvements. And, with oil capacity in surplus in each Gulf state, it could have the benefit of polarizing the fragmentation, thus putting every indigenous state in the race with every other for an increased and uninterrupted flow of oil.

A major Soviet assault on the Gulf states is not, of course, the responsibility of the 6,000 soldiers.[12] A development of this sort would call into action another superpower whose interest is even greater than that of the British, i.e., the United States.

With this in mind, the British saw no value in continuing their military presence. In fact, some said that the British military presence was bound to incite nationalists and extremists into hostile actions against Western installations, whereas the original intention was the prevention of such hostility.[13] Still others referred to the Popular Front for the Liberation of the Occupied Arab Gulf (PFLOAG), a revolutionary movement based in Dhufar and given training and military aid by the Chinese and Russians, as the most menacing movement against Western presence.[14] British departure, then, was said to have been prompted by the British fear of being snared 'in a Vietnam-like trap.'[15]

Although one cannot disregard these contingencies as hastening the British withdrawal, one can nevertheless see that these developments do not seriously present themselves. According to T.B. Millar, 'There is just enough truth, in these propositions in general to make them almost wholly misleading in particular cases.'[16] Western presence and the oil installations are almost entirely confined to a vast and open desert which is, necessarily, sparsely populated, a climate wholly uninviting to successful guerrilla warfare. Besides, the population of Oman, where the PFLOAG is said to have its base, is put at half a million people – a situation in which a small and well organized police force could round up the whole population in a matter of days.

125

To turn now to the major crises said to be contingent on the British withdrawal are (1) the rivalry – territorial or otherwise – among the indigenous states, (2) the ensuing instability and disorder, and (3) the impact of the first two on the continuous flow of oil.

The forces of rivalries and disorders are conspicuously present. Yet, no other crisis had better convinced the British of the irrelevancy of their military presence than the one fomented by Dr Musaddeq in Iran in 1951. 'It is certainly true,' says Sir William Luce, 'that British troops neither can nor should protect the oil industry in any direct sense.' Sir William continued: 'Even in the extreme case of nationalization of British oil interests, should this ever arise, there should be no question of forcible intervention.'[17] This being established, the British further learned that it is within the context of this rivalry that the Iranian oil nationalization of 1951 failed, and the other Arab states made up for the Iranian interruption. If further proof of the effect of the built-in rivalry is needed, look to the development of the Arab–Israeli War of 1967. The British military forces could not, then, stop the Arabs from boycotting the West. However, the shortage in oil production was then met by the Iranians.[18]

As a result of these developments, the British were progressively disposed to the idea of ending their military presence. At the same time, they were busy laying the foundation of a policy of built-in conflicts – the policy of 'balanced tensions and conflicting claims,' to reiterate the words of D.C. Watt. Following is an elaboration of this policy.

The Arabian Peninsula is a vast and waterless desert with no real boundaries as the West knows them. Except for a few oases, there are hardly any trees. It is a barren and uninviting land with little motivation among its inhabitants to mark its boundaries. The scarcity of rainfall forced upon its dispersed tribes a seasonal wandering after the grass – an activity which made a farce of any attempts at border delineation.[19]

This was the situation before the British 'froze' tribal wanderings and obstructed the emergence of a central force similar to the one that emerged in Saudi Arabia. If there had been no

Britishers with 'the right to fix State boundaries and to settle disputes between the Trucial Sheikdoms' in accordance with the treaties of 1820, 1835, 1853, and 1892, there would, at least, not be a barren coastal land of 300 miles, seven city-states, with every sort of boundary and dependency one could think of.[20]

There would not be a State of Ajman enclaved by Sharjah which is in conflict with Umm al-Qaiwain which borders Ras al-Khaima whose domain is intercepted by Fujaira whose jurisdiction is divided by a Sharjah Dependency whose border touches on a Dependency of Dubai whose Sheikh is in conflict with the Sheikh of Abu Dhabi whose domain, together with that of the Trucial States, divides that of Oman, which together with Abu Dhabi are disputing the jurisdiction of the Buraimi oasis with Saudi Arabia.[21] This paragraph is decidedly confusing and one is referred to the map drawn by the British to see if he can learn better.

The list of ill-defined boundaries, border confusions, territorial claims and counter-claims is just formidable. Britain seems experienced in dividing nations and, inevitably, as in the case of Sharjah in 1866, families. To see the British at work, a general clarification is provided by Richard Coke, who started his analysis by raising the question: 'Has the Arab, then, gained or lost by the added closeness of the British connection?' His answer expressed mixed feelings. He said:

> In the material sense, he has undoubtedly gained ... In the political sphere the Arab has gained less. A cynic, perhaps, might even suggest that the muddle which the Allies created in the Arab lands by their post-war settlement outweighs the good their conquest brought about in non-political ways.[22]

However, Coke is puzzled by the blessing said to have been brought about by the British. He thus qualified the previous statement and admitted:

> Even commercial development has, it must be admitted, been seriously affected by political issues. The oil questions, the passport nuisance, the varying currencies, official languages,

import and export duties are all of them, if not political in origin, deeply influence by political considerations.[23]

Coke offered some sympathy for the Arab catastrophe. He is, however, a Britisher; and the sympathy was offered in deprecation so as to rob an Arab of the benefit of a reader's seeing what the British had done and wishing them, at least, 'bad luck.' He said:

> It is easy to sympathise with the vexation of the Arab
> commercial man who, having been brought up under the
> Turkish Empire in which free progress for men and goods
> between the various Arab cities was interrupted only by
> brigands, now finds across his path a variety of boundaries
> and frontiers which might even tax the patience of the
> Balkans.[24]

According to Coke British muddling did not limit itself to commercial activities only. He said:

> Unfortunately, the resulting complication does not stop at
> passports, weights and measures; it pervades the whole
> atmosphere of various administrations, which differ radically
> from each other in organization, in method, and in aim.
> Democratic government, for example, has been actively
> encouraged by the British in Iraq; it had been actively
> discouraged by the same British in Palestine.[25]

Coke's analysis concluded with a question: 'Is it surprising that the Arab is sometimes puzzled by this entire lack of coordination of plan and idea?' and answered thus: 'A uniform system of control, even when badly applied as in the case of Turkey, is at least comprehensible; you may not love it, but you can soon settle down to its working.'[26]

Inclusion of these citations has been for the purpose of showing that contrary to what is often said, the British were consciously pursuing a policy of creating tensions and conflicting claims. Their 'recommendations' were in the form of orders.[27] They deposed leaders, sheikhs, and sultans; they divided domains, and they perfected the fragmentation. In 1956, five Bahraini leaders were deported for their demand of 'a more popular form of government.' In 1965, Sheikh Sagr of Sharjah was deposed despite an appeal to the United Nations Secretary-General, followed by Sheikh Shakhbut of Abu Dhabi, who was ousted in 1966. On July 24, 1970, the Sultan of Oman, Said Bin Taimour, was deported.[28]

In 1856, on the death of the Sultan of Oman, the British divided his domain between his two sons, and in 1866, on the death of the Sheikh of Sharjah, the tiny Sheikhdom was divided by the British among his four sons into Sharjah, Ras al-Khaima, Dibah, and Kalba.[29]

After perfecting the fragmentation and creating so many entities, each with its jealousies and peculiarities, the British gave, in March 1968, their permission for the formation of a 'federation of Arabian Emirates' among Bahrain, Qatar, and the seven sheikhdoms of the Omani Coast. From its inception the federation has been so entangled with problem of vested interests that, in the end, Bahrain was prompted to declare its independence and became a sovereign member of the United Nations in September, 1971. Bahrain was then followed by Qatar. The other seven 'city-states' then joined together in July 1971, to form their loose federation, only to be boycotted by the Sheikh of the State of Ras al Khaima, who decided to stay aloof.[30] One should not be surprised at the failure of the Federation because the 'British presence in the Persian Gulf forced an artificial area character upon the separate states – one which properly would not have arisen spontaneously.'[31]

A similar situation developed in another part of the Arabian Peninsula when the British captured the Sultanate of Lahej only to draw a boundary between the Sultanate and its port, Aden. Having drawn the boundary, the British protectorate of Lahej was left to suffer from its extreme material backwardness, while its port, Aden, was put under a complex colonial administration. In the process, there developed two distinct ways of life which made it difficult for the Adenese Arabs to accept the Lahejese Arabs or for the Lahejese to accept the Adenese.[32]

Along the way to intensifying the division between the body and its members, the Arabs of the Lahej protectorate, although working in Aden, were denied participation in the political system of Aden. Thus, while the Indians, the Europeans, British and non-British, and the Jews were considered part of Aden society and therefore allowed to participate in the political system, the Lahejese were denied the right to vote. According to the British Governor of Aden, the Lahejese were 'unfit' to exercise political power and therefore 'the electoral net' was 'spread with care.' The British Governor, moreover, said this after 150 years of British occupation.[33]

These are but a few examples of British muddling. Further north, along the Arabian shore of the Gulf, one is struck by the presence of 'unusual features' of two Neutral Zones between Saudi Arabia and Kuwait, and Saudi Arabia and Iraq.[34] The carving of these two zones was the work of Sir Percy Cox, the British Political Resident, who was then representing Kuwait and Iraq in the negotiations with King Abdalaziz Ibn Saud of Saudi Arabia.[35] The two parties could not reach agreement on the boundary between Saudi Arabia and Iraq at the conclusion of the Muhammarah Treaty on May 5, 1922.[36] Seven months later, December 2, 1922, the parties gathered at al-Ugair, a Saudi port on the Persian Gulf, where King Abdalaziz, representing Saudi Arabia, and Sir Percy Cox, representing Iraq and Kuwait, agreed on the delineation of the borderlines between Iraq and Saudi Arabia and Kuwait and Saudi Arabia. The agreement with Iraq was incorporated in a protocol to the Treaty of Muhammarah and the agreement with Kuwait was covered by the Treaty of al-Ugair. The emergence of the two Neutral Zones was the result of these two treaties.[37]

Conflict is, of course, what one should expect from such arrangements, for who is to have administrative control over these neutral zones and which states should have the first right in chartering their exploitation, especially in an area known for its prolific oil resources? Luckily, the authorities in both Saudi Arabia and Kuwait were able to contain their dispute over their neutral zone by dividing the zone into two equal administrative spheres while reserving the zone's natural resources to be exploited by the two states on an equal basis. No arrangement of this sort has been concluded between Iraq and Saudi Arabia over the neutral zone between their states.

Now Iran. The only frontier Iran has with any Arab state is with Iraq. And here the British Government 'engineered' the most unusual border between any two states.[38] Near the Gulf, Shatt al-Arab, a waterway of about 120 miles formed by the confluence of the Tigris and Euphrates, is the international dividing line between Iraq and Iran. The border on this waterway is governed by the terms of the treaty of 1937 between the two states. But, instead of delineating the border between the two countries on the median of the waterway, as is usually done in accordance with the Thalweg principle, the 1937 treaty gave Iraq, which was then controlled by the British, sovereign power over the whole waterway – up to the eastern bank of the Shatt.[39]

With this boundary arrangement, one would expect to find a port, similar to the Iranian port of Abadan with all its complexity and a refinery (one of the largest in the world) on a shore, yet that shore is the sovereignty of another state. Ships navigating the Shatt al-Arab and bound for Abadan have to fly the Iraqi national flag and Iranian fishermen have to have permission from Iraq to fish in the Shatt.

Now, irrespective of the potential presence of oil under the Shatt al-Arab, a situation of this sort is bound to touch on the pride of the Iranians and stir their national feelings. To have an inland state without a seacoast is already a cumbersome situation usually attributable to bad luck. But to have a port without a shore is outrageous and can only be, as the Iranians like to stigmatize the arrangements, 'a vestige of imperialism.'[40]

The Shatt al-Arab has, therefore, been a source of continuous friction between the Iranian and Iraqi governments. In 1966 the two governments agreed to start a discussion on the problem only to have it postponed.[41] A discussion would, of course, put Iraq at a disadvantage. It would necessarily lead to a rectification of the navigation rights incorporated in the treaty of

1937. For Iraq, this would amount to a renunciation of her sovereignty over a part of her territory – an act which no Iraq government could do without losing its popularity. The dispute thus erupted in April 1969, into a situation of amassing troops on both sides of the border. In its attempts to force negotiation, Iran decided to assert a *de facto* abrogation of the 1937 treaty by sending her ships through the Shatt al-Arab waterway flying the Iranian national flag and guarded by heavy naval escort. Iraq protested the use of force. The problem, however, is still unsettled.

Another example of the mechanism of the built-in dispute at work is Iran's capturing of three little islands – Abu Musa, Big Tunb, and Little Tunb – near the Strait of Hormuz. The British had always maintained that the island of Abu Musa belongs to the Sheikhdom of Sharjah and that Big Tunb and Little Tunb belong to the Sheikhdom of Ras al Khaima. The rulers of both sheikhdoms were exercising political control over their respective islands through the British direction. Meanwhile, Iran was laying claim over these islands when the British presence came to an end without a final settlement of the dispute. Before the British left, however, Iranian forces occupied the islands on November 30, 1971. The occupation caused some bloodshed and stirred a diplomatic crisis among Iran and Iraq and some of the Arab states.[42]

The crisis precipitated the expulsion of thousands of Iranian nationals from Iraq and soured the Arab-Iranian relationship. A reaction pertinent to the oil question was the stand taken by Libya and Algeria. Both countries declared their intention of abstaining from attending the conference of the Organization of Petroleum Exporting Countries (OPEC) – the only instrument that organizes them into a united front to face the oil companies' bloc – held in Abu Dhabi in December 1971.[43]

What the British fear most is concerted action on the part of the oil-producing countries. If such action were to take place, the 6,000 soldiers stationed on the island of Bahrain would be unable to prevent it. After all, the British military forces could not prevent the achievements realized by the Organization of Petroleum Exporting Countries (OPEC). Thus, on every account the British military presence was proving unworthy of the financial burden – though very minimal, estimated at £17 million a year[44] – compared to the profits the British are reaping. They thus decided to leave; but not before they finished sowing the seeds for the causes and forces of tensions and fragmentations – causes and forces which would prevent unified action on the part of the oil producing countries. That accomplished, a built-in dispute is thus perpetuated and a complete oil interruption is reasonably eliminated.

The Surplus Capacity

Iran came out of her crisis with all kinds of social and economic disorders. The message brought home to the Iranians and the Arabs was that 'doing a Musaddeq hardly pays off.'[45] According to Herbert Feis the lesson for Iran might not be all that bad. For these local regimes must learn not to put a price on their friendship and cooperation. 'We,' continued Dr Feis, 'ought not to expose our security to the possible mismanagement, missteps, or ill will of the regimes in these various Moslem countries.'[46]

One could conclude from this point of view that oil controlled by a country, and in whatever other country the oil may be, is to be considered a matter related to the national security of the controlling country. As such, a controlling country may resort to the hindrance of the development of the oil resources of a producing country if the latter action is not in line with the requirements of the former's national security. Feis put this in the negative and stated: 'We ought not to hinder the development of their oil resources except for clear reasons of broad national welfare or security.'[47]

The question which now arises is how was it not possible for Iran to resist the hindrance of the development of its oil resources, or, to put it differently, how, despite the strategic significance of oil, could the oil consuming countries overcome the stoppage of the Iranian oil and, thereby cause the collapse of the Iranian nationalization? Similarly, but to a lesser degree, how were the consuming countries able to overcome the Arab oil boycott after the Arab–Israeli war of 1967?

Gilbert Burck provides some answers to these questions. Still in the context of the control of the oil by the major oil companies, Burck stated:

> The big oil companies, because they all have large *alternative* sources of supply in the Middle East, are not dependent on any one country. So long as they control refining and marketing, they theoretically can still discipline a country that asks for too much – just as they disciplined Musaddeq.[48] (Emphasis added.)

A first impression on the effectiveness of the surplus capacity (alternative source of supply) is thus provided.

Oil in the Persian Gulf, as we know, is almost 100 percent foreign owned[49] and the great bulk of the oil business is carried by the seven international companies – also referred to as the majors, cartel, concessionaire, and consortium. These are Standard Oil Company (New Jersey), Standard Oil of California, Texas Oil Company, Mobil Oil, Gulf Oil (Americans); British Petroleum (British); Royal Dutch-Shell Group (British-Dutch). Ownership in itself is an important factor in the control of oil. The crucial leverage in the effectiveness of the surplus capacity, however, is the way in which ownership is organized. The industry is jointly owned and geographically diversified. Every single one of the majors is, in one way or another, a partner of the other majors, and every major has at least two sources of supply in not fewer than two countries of the Gulf area. (See Table 11.)

The diversity of the geographical sources of supply has been sought by the companies for economic reasons, but also for security purposes – to avoid being dependent on a limited number of supply points. The importance of the geographical diversity of supply sources is, with oil capacity in surplus, to provide the companies with a choice of where to produce the oil they need. The importance of such a choice in the hands of the companies can hardly be exaggerated. It means that if any one oil-producing country asks for more, that country may be asked to 'drink it' with more oil pumped up from its neighbor.

The effectiveness of this instrument depends, of course, on the kind of relationship prevailing between the indigenous oil-producing countries themselves. But we also know that before this type of company organizational setup stands a group of quarreling states whose dependence on oil for their revenues reaches an unmanageable level – in most of the cases revenues from oil reach over 80 percent of the total government receipts.

The consortium in Iran is almost totally controlled by the seven majors. Despite the efforts made by the Iranians to interest independent companies, 90 percent of 1970 Iranian production was contributed by the consortium. In 1970 consortium production amounted to 3.5 million bpd against an average production of 3.8 million bpd for a final total production for the same year.[50] Accordingly, the government revenues derived from the consortium form by far the greatest part of the foreign exchange derived by Iran. In 1968 the consortium payment amounted to £338 million. Against this amount other companies made the payment of £17 million for the same year.[51]

In Iraq the situation is not different. The oil operations are almost totally dominated by the Iraq Petroleum Company (IPC) and its associated companies, the Basra Petroleum Company and the Mosul Petroleum Company. Oil production in 1970 amounted to 75 million tons. Total government revenue for 1969-70 was put at £408 million, of which £200 million, or around 50 percent, was made up of contributions from oil.[52]

The oil business in Kuwait is dominated by the Kuwait Oil Company, a subsidiary of Gulf Oil Corporation and British Petroleum. Kuwait Oil Company was responsible for 87 percent of the oil production, which in 1969 amounted to 139 million tons. Estimated government revenue for 1970 was put at £355 million, of which £326.9 million, or 93 percent, was provided by the oil industry.[53]

The Arabian American Oil Company (Aramco) in Saudi Arabia is controlled by four of the major oil companies. Oil production in Saudi Arabia is largely dominated by Aramco. Out of the 1,024 million barrels of oil produced in 1967 Aramco's contribution was 948 million barrels, or 93 percent of the total Saudi Arabian production. Similarly, in 1969, Aramco paid $895 million to Saudi Arabia against $52 million made by other concerns engaged in the oil business in Saudi Arabia. As expected, a very large portion of Saudi Arabia's revenue comes from oil operations, which in the fiscal year 1970-71 reached 87 percent.[54]

Similar dependence on revenues from oil is to be expected in Qatar, Bahrain, and the Trucial States. These states have practically no source of income other than the extraction of oil or its related activities.

This statistical information is provided to underscore the economic situation which is largely dominated by the oil business. When this economic picture is further expanded to include the geographical diversity of the supply sources which gave the companies the ability to have a choice of where to produce the oil they need, it is not difficult to comprehend the kind of instrument the companies possess in controlling the policies of the producing countries. A general description of the effectiveness of this instrument is provided by Dr Feis:

> Perhaps even more crucially, the foreign exchange provided by oil exports has become essential for the maintenance of the economy of the country. This means in effect that any and all would suffer severely, as Iran has, from any substantial decline in oil production and exports.[55]

The prospect of Iranian nationalization was brought to its end because the loss of 650,000 bpd of crude oil was made up for by increases in the production of Iraq, Kuwait and Saudi Arabia. During the period of 1950-53 Iraq's oil production increased from 136,200 bpd to 576,000; Kuwait's, from 345,000 bpd to 861,700; and Saudi Arabia's, from 546,700 bpd to 844,600. 'Because of this unprecedented rise in Middle East oil production from Kuwait, Iraq, and Saudi Arabia, the world did not come knocking on National Iranian's door for oil, as had been expected.'[56]

Similarly, the Arab boycott of some Western states after the Arab–Israeli war of 1967 was mitigated because Iranian oil production was stepped up. The consortium then filled the gap created by the Arab embargoes. In 1967, Iranian oil production rose sharply, by 22 percent. Consortium production went up from 98.8 million tons in 1966 to 120.9 million tons in 1967.[57] Again, because of the unprecedented increase in Iranian oil production the Arab boycott went virtually unnoticed.

It is noteworthy in concluding this section that the effectiveness of the instrument of the surplus capacity has been placed within manageable proportion with the emergence of the Organization of the Petroleum Exporting Countries and the prevailing of a general understanding among the oil-producing countries that their interest lies in their harmony. It remains to be seen, however, whether these countries will refrain from being tempted to play one against the other.

Endnotes

[1] The Center for Strategic and International Studies, *The Gulf: Implications of British Withdrawal*, Special Report, Series No. 8 (Washington DC: Georgetown University, 1969), 64.

[2] Ibid.

[3] Anthony Verrier, 'The British Withdrawal from the Gulf and Its Consequences,' *The Princeton University Conference and Twentieth Annual Near East Conference on Middle East Focus: The Persian Gulf, October 24-25, 1968*, Cuyler Young (ed.) (Princeton: The Princeton University Conference, 1968), 136.

[4] Robert G. Landen, 'The Modernization of the Persian Gulf: The Period of British Dominance,' *The Princeton University Conference and Twentieth Annual Near East Conference on Middle East Focus: The Persian Gulf, October 24-25, 1968*, Cuyler Young (ed.). (Princeton: The Princeton University Conference, 1968), 20; David H. Finnie, *The Middle East Oil Industry in its Local Environment* (Cambridge: Harvard University Press, 1968), 2.

[5] *The Middle East and North Africa, 1970-71* (17th edn.; London: Europa Publications Limited, 1970), 574; Halford L. Hoskins, 'Changing the Guard in the Middle East,' *Current History*, LII, No. 306 (February, 1967), 70.

[6] *The Gulf: Implications of British Withdrawal*, op. cit., 4; T.B. Millar, 'Soviet Policies South and East of Suez,' *Foreign Affairs*, XLIX, No. 1 (October, 1970), 79.

[7] Roy E. Thoman, 'The Persian Gulf Region,' *Current History*, LX, No. 353 (January, 1971), 45.

[8] 'Britain Will Withdraw from the Gulf,' *The Middle East Economic Digest*, XIV, No. 15 (December 18, 1970), 1470; 'Persian Gulf Vacuums,' *Christian Science Monitor* (March 4, 1971), 14.

[9] William Luce, 'Britain in the Persian Gulf: Mistaken Timing Over Aden,' *Round Table*, LVII (July, 1967), 260; Godfrey Jansen, 'Crisis Potential in the Persian Gulf,' *Christian Science Monitor* (October 29, 1970), 2.

[10] D.C. Watt, 'Britain and the Future of the Persian Gulf States,' *World Today*, XX (November, 1964), 489; Ravinder Kumar, *India and the Persian Gulf Region 1858-1907: A Study in British Imperial Policy* (New York: Asia Publishing House, 1965), 211.

[11] John Allan May, 'Russians in the Indian Ocean,' *Christian Science Monitor* (November 17, 1970), 13.

[12] Finnie, op. cit., 2; Thoman, op. cit., 44.

[13] Thomas, ibid., 353.

[14] See David Holden, 'The Persian Gulf: After the British Raj,' *Foreign Affairs*, XLIX, No. 4 (July, 1971), 727-728.

[15] 'Russia Drives East of Suez,' *Newsweek* (January 18, 1971), 29.

[16] Millar, op. cit., 79.

[17] Luce, op. cit., 280.

[18] Holden, op. cit., 729; Hoskins, op. cit., 70.

[19] Roderic Owen, *The Golden Bubble: Arabian Gulf Documentary* (London and Glasgow: Collins Clear-Type Press, 1957), 141-142; Richard H. Sanger, *The Arabian Peninsula* (Ithaca and New York: Cornell University Press, 1954), 152; Sir Rupert Hay, *The Persian Gulf States* (Washington DC: The Middle East Institute, 1959), 114.

[20] Holden, op. cit., 732; *The Middle East and North Africa, 1970-71*, op. cit., 586; John S. Badeau, 'Inter-Arab Social and

Political Relationships,' *The Princeton University Conference and Twentieth Annual Near East Conference on Middle East Focus: The Persian Gulf, October 24-25, 1968*, Cuyler Young (ed.), (Princeton: The Princeton University Conference, 1968), 195.

[21] Sanger, op. cit., 176; J.B. Kelly, 'The British Position in the Persian Gulf,' *World Today*, XX (June, 1964), 245; Watt, op. cit., 489-490; *The Middle East and North Africa, 1970-71*, op. cit., 586-588; *The Economist*, June 6, 1970, 72; Holden, op. cit., 726-727.

[22] Richard Coke, *The Arab's Place in the Sun* (London: Thornton Butterworth Ltd., 1929) 295-297.

[23] Ibid.

[24] Ibid.

[25] Ibid.

[26] Ibid.

[27] Badeau, op. cit., 195; Watt, op. cit., 489; *The Economist* (June 6, 1970), 72.

[28] Hay, op. cit., 93; *The Middle East and North Africa, 1970-71*, op. cit., 587.

[29] *The Middle East and North Africa, 1970-71*, op. cit., 557 and 587.

[30] *The Middle East and North Africa 1971-72* (18th edn.; London: Europa Publications Limited, 1971), 207; United Nations, *Monthly Chronicle*, VIII, No. 10 (November, 1971), 70; 'Control of the Gulf,' *Christian Science Monitor* (December 8, 1971), 8.

[31] Badeau, op. cit., 195; Landen, op. cit., 7.

[32] A.S. Bujra, 'Urban Elites and Colonialism: The Nationalist Elites of Aden and South Arabia,' *Middle Eastern Studies*, VI, No. 2 (May, 1970), 193.

[33] See ibid., 197-198 and 203.

[34] *The Middle East and North Africa, 1969-70* (16th edn.; London: Europa Publications Limited, 1969), 285.

[35] David Howarth, *The Desert King: Ibn Saud and His Arabia* (New York, Toronto, London: McGraw-Hill Book Company, 1964), 139; Hay, op. cit., 98.

[36] J.B. Kelly, *Eastern Arabian Frontiers* (London: Faber and Faber, 1964), 113.

[37] Hay, op. cit., 98; Howarth, op. cit., 139.

[38] *The Middle East and North Africa, 1969-70*, op. cit., 285 and 295.

[39] Thoman, op. cit., 45; *The Middle East and North Africa, 1969-70*, op. cit., 256 and 291.

[40] Thoman, op. cit., 45.

[41] *The Middle East and North Africa, 1969-70*, op. cit. 255.

[42] John K. Cooley, 'It's Several Tails Wagging Several Dogs in Middle East,' *Christian Science Monitor* (December 4, 1971), 7; 'Control of the Gulf,' *Christian Science Monitor* (December 8, 1971), 18.

[43] John K. Cooley, 'Iraq Charged With Mass Deportation of Iranians,' *Christian Science Monitor* (January 8, 1972), 4; *Al-Ahram* (December 8, 1971), 1.

[44] Holden, op. cit., 725.

[45] Wanda Jablonski, 'Master Stroke in Iran,' *Collier's* (January 21, 1955), 34.

[46] Herbert Feis, 'Oil for Peace or War,' *Foreign Affairs*, XXXII, No. 3 (April, 1954), 424.

[47] Ibid., 425.

[48] Gilbert Burck, 'World Oil: The Game Gets Rough,' *Fortune*, LVII, No. 3 (May, 1958), 188.

[49] At the time of this writing, it was brought up in the news that an agreement had been reached between Saudi Arabia and Aramco under which the company agreed to a direct 20 percent Saudi government ownership in its operations. The agreement is expected to have far-reaching impact on the changing relationship between international oil companies and the oil-producing countries. The topic is fully discussed in the chapter on the Organization of Petroleum Exporting Countries (OPEC).

[50] *The Middle East and North Africa, 1971-72*, op. cit. 239.

[51] Ibid., 238.

[52] Ibid., 281, 283 and 286.

[53] Ibid., 346, 398 and 402.

[54] *The Middle East and North Africa, 1969-70*, op. cit., 584; *The Middle East and North Africa, 1971-72*, op. cit., 514.

[55] Feis, op. cit., 424.

[56] Charles W. Hamilton, *Americans and Oil in the Middle East* (Houston, Texas: Gulf Publishing Company, 1962), 59.

[57] *The Middle East and North Africa, 1969-70*, op. cit., 259; 'Agreement Reached in Iranian Demands for Another Year,' *World Petroleum*, XL, No. 7 (July 1969), 20A.

6

The Soviet Union: A Renewal of the Power Struggle for the Control of Oil

The highest concentration of American private investment in any comparable region is found in the Persian Gulf area, according to William D. Brewer of the US Department of State. Testifying before the Subcommittee on the Near East of the United States House of Representatives' Committee on Foreign Affairs, Charles Bonin, President of the American-Arab Association for Commerce and Industry, stated on July 29, 1970, that the book value of all United States direct investment in the Arab countries was estimated at S2,200 million in 1969, of which over 90 percent was taken by the petroleum industry.[1]

Net annual remittances made to the United States by American oil companies operating in the Persian Gulf area were put at over $750 million. Another estimate put the percentage of income derived from the Middle East (including Iran but excluding Libya) in 1968 at about 15 percent of all the United States investment income derived from abroad. Bonin calculated the figure from the entire Arab world at $1,400 million or 28 percent of the total 1968 income derived from all US direct investment abroad.[2]

An ever-increasing portion of the taxes and royalties paid to the oil-producing countries finds its way back to the United States in exchange for American goods and services. According to Brewer, United States exports to the Persian Gulf region amounted to $137.3 million in 1947, went up to $314.6 million in 1957, and reached $600 million in 1967. Trade with all Arab states has averaged S1,000 million in recent years.[3]

In contrast, United States imports from the Persian Gulf region were put at $34.2 million in 1947. The figure went up to $227.6 million in 1957 but then declined to $188.8 million in 1967.

This netted the United States $411.2 million in trade surplus in 1967. Bonin put the trade surplus in favor of the United States as a result of her trade with all Arab States in 1969 at $545,000,000, or 28 percent of America's surplus all over the world.[4]

In addition, the Gulf region assisted the United States balance of payments position through the purchase of long-term United States securities. Annual Arab capital inflow to the United States was conservatively estimated at $200 million.[5]

The Center for Strategic and International Studies estimated the Gulf region's annual contributions to the United States and British balances of payments at between $1.5 and $2 billion. The Center then made the following impressive statement:

> It should be emphasized in this regard that the security of the West cannot be separated from its fiscal condition – especially the balance of payments problem.[6]

The preceding data are offered to give a sense of the economic stake of the United States in the Persian Gulf area. The Persian Gulf, moreover, is, as we know, a tiny arm of the Indian Ocean – the busiest sea route in the world. On its routes pass 50 percent of Europe's oil, 90 percent of Japan's, 60 percent of Australia's and 80 percent of Africa's, all originating from the Persian Gulf area.[7] The US Department of the Interior estimated the flow of oil in international trade in 1969 at 21.8 million barrels per day. Nearly 11 million barrels per day originated from the Persian Gulf region, of which 9.3 million, or 86 percent, were shipped by tankers through the Strait of Hormuz and into the Indian Ocean.[8]

The number of ships passing through the eastern entrance of the Indian Ocean, the Malacca Strait, including oil tankers average 150 ships per day. The number of ships using the sea lane around the Cape of Good Hope, the western entrance of the ocean, was put at 24,000 ships per year.[9]

In view of this United States economic involvement, the massive Indian Ocean traffic, the nature and significance of the commodity transported, and its impact on Western balances of payments, prosperity and security, many were greatly disturbed at the virtual disappearance of Western forces, particularly American, from the Gulf area and the Indian Ocean in general. With the British withdrawal from the Persian Gulf area complete, there developed a vacuum 'so much nearer Moscow than Washington' and over which the Soviets are tempted to assert their strength.[10]

Today, the Soviet Union has between 15 and 30 warships patrolling the Indian Ocean. Among these are guided missile vessels, destroyers and submarines. Recently, the Russians got an agreement with the island of Mauritius for aircraft landing rights and the establishment of a base for trawlers. They asked the government of Ceylon for a base at the port of Trincomalee and negotiated with the government of India for rights to help them in their naval activities. Soviet activities in the Indian Ocean range from fishing and communication to satellite tracking, missile testing, and deployment of naval power.[11]

If one is to follow the Soviet track, the Soviets now appear well anchored in the Mediterranean, on the Suez Canal, at Aden, and in the Indian Ocean – a course of action which 'stems from geopolitical reasons not dissimilar to those that first induced Britain' to set up bases during the heyday of the British empire.[12]

Despite this conspicuous Soviet advance in an area of great economic interest to the West and to the United States in particular, United States military presence is almost non-existent. It is confined to a seaplane tender, listed as a flagship, and two old destroyers based on the island of Bahrain in the Persian Gulf. According to Professor T.B. Millar, an international authority on the defense of Australia, 'The United States has given no indication of intending to replace Britain or compete with the Soviet Union in the Indian Ocean.'[13] The United States has lately concluded an agreement with Bahrain for the continuation of her naval base on the island. However, that agreement was prompted by a legalization of a presence after the British departure, rather than a move toward a drastic change in the United States policy in the Persian Gulf.

A number of explanations are offered to justify America's coolness to this renewed Soviet drive. One such explanation is offered by Millar under the concept of the spheres of influence arrangements. Millar interpreted American statements and actions 'as implying an acknowledgement that they are prepared to concede in some parts of the area a superior Soviet influence; in other parts they may seek to coexist.'[14]

The problem with the spheres of influence arrangements, however, centers on the point of where to draw the line between the contestants. The moment a sphere of influence is established, the establishing power still looks for another sphere of influence to enhance its security. At any rate, precedents support this contention. From being a completely Western bastion, the Russians had, in a matter of 15 years, put into their sphere of influence most of the countries of the area. Today, for example, the Soviet Union is a power with decided influence in Egypt, Syria, Iraq, and South Yemen. She was instrumental in helping Iran, Libya, and Algeria break away from Western oil control and, despite a recent setback, she is more influential in the Sudan and in the Arab world in general, at least until the Palestinian problem is settled.

Another explanation for America's indifference to the Soviet presence in the Persian Gulf and the Indian Ocean is founded on the global strategy of the United States, itself based on the theory of an assured deterrence capacity. Now that the United States has established legal rights over the oil resources of the Persian Gulf area, she would not expect any direct Soviet interference with the installations and line of communications, and any such interference might risk big-power confrontation.[15] The deterrence concept further assumes that if a military confrontation between the superpowers is unavoidable, it is more efficiently launched from the Atlantic and Pacific or the Mediterranean, with the Indian Ocean remaining as a backwater of less importance in the United States global strategy.[16]

This is indeed a happy reliance on a policy of either nuclear confrontation or do-nothing. It is an oversimplification, because, for example, the Russians made substantial inroads into the Middle East without even firing a single shot.

Is it really necessary for the Soviet Union to resort to total war to gain control over any or all of the Persian Gulf states? The Soviet approach has consistently been one of prudent caution. It is inconceivable, therefore, that the Soviet Union would make an overt commitment to a country like Iraq or Iran so as to make the United States defend what it considers its strategic interests, an act conducive to total war. Lincoln Landis, senior research fellow at the Georgetown Center for Strategic Studies, would say that since the Soviet Union boasts of vast oil discoveries, her recent drive in the direction of the Persian Gulf is 'indicative of heavy political overtones.' Landis has not ruled out the development of a situation in which the Soviets would make an 'overt commitment' to any of the oil-producing countries of the Gulf area. In Landis' opinion this is 'the beginning of a new long-range Soviet resolve to confront the West in this region.'[17]

Can it be expected that the Soviet demand for foreign oil would be considerable enough in the near future so as to make a direct Soviet action against the states of the Persian Gulf and the ensuing worsening of American-Soviet relations palatable to American strategists? Robert E. Hunter said that in such an event 'direct Soviet action against states in the Persian Gulf area cannot be entirely ruled out.' He qualified his statement by saying that direct Soviet action 'is highly improbable.'[18] Obviously this qualification was offered with the deterrence theory in mind.

Nevertheless, the deterrence concept was challenged on other grounds. As the Soviet Union reaches strategic parity with the United States, how likely is it that the United States will be ready to resort to nuclear confrontation in defense of some interests in some distant corner of the world and thereby inflict an 'unacceptable' destruction on herself? According to Paul Cohen, a 'threat of nuclear war could be invoked with limited danger to [the US] when a grossly unequal balance existed between America and foreign nuclear power.'[19]

Within the context of discussing submarine supremacy in modern naval warfare – an obviously related topic to the oceanic communications discussed in the preceding paragraphs – Cohen made the following revealing suppositions. He said:

> ... suppose that an enemy decides to cut our maritime
> communications in some selected sector of the world, or over

all the oceans ... Suppose that the enemy has carefully
refrained from using nuclear weapons, or from threatening
American territory. What he wants is a free hand abroad, in
some area in which his internal lines of communications can
suffice. (For he can no better send his merchant ships through
our submarine blockade than can we send ours through his.)
In our chagrin, in our shame at abandoning our allies, in our
fear of living in dingy poverty on what our own land can
supply, would we then call for mutual suicide by attacking
the enemy with nuclear weapons?[20]

Cohen chose to answer the question and the suppositions in this
manner: 'Even our Allies might not appreciate the gesture.'[21] He
substantiated his contention with cases from recent history. He
said: 'If the stakes are less than total, if they involve, in some
distant corner of the world, a blow to our pride or a loss to our
pocketbook, will we not be forced to the same restraint we have
shown in Korea and Vietnam?'[22]

The renewed Soviet drive in the Persian Gulf has thus gone
unassessed. Western withdrawal having been completed with the
British departure, it is not difficult for the Soviet Union to find the
techniques and justification, short of war, to increase her influence
in an area just across the border. Thus, while an atmosphere of
indifference and coolness is prevailing in the United States attitude
towards its interests in the Persian Gulf area and the Indian Ocean
in general, the Soviets are busy taking 'advantage of Western
weakness whenever it occurs, both to further specific Soviet
interests and to improve the Soviet–Western ratio.'[23]

The Soviets seem to be embarking on a three-dimensional
policy with regard to oil resources in the Persian Gulf area. The
first is to gain control over the oil-producing countries, or, short of
that, to open new options to these countries, helping them to break
away from Western control – the 'spoiler's' policy, as it is usually
called. The second is to compete with American and British
companies in the energy market of Western Europe, earning a
handsome sum of foreign exchange and further helping in the
acceleration of the waning of American influence. The third

dimension of their policy is to tighten their control over Eastern European countries through the control of their energy supply. Following is an elaboration of these points.

With all its attendant contradictions, pragmatism (or the policy of expediency), rather than ideological fervor, has been the motivating force behind Soviet foreign policy. Thus, while the Palestinians were compromised with the Soviet-American Middle Eastern détente and Walter Ulbricht paid the price for Moscow-Bonn reconciliation, the Soviets engaged in a war of phraseology that is always in direct proportion to the firmness of the decision to take no action.[24]

Within the context of this pragmatic approach is the recent Soviet drive to conclude Western-type, business-like, long-term contracts with the oil-producing countries of the Persian Gulf. The Russians were laying pipelines, exploring for and developing oil fields, and transporting and marketing crude oil and oil products into Eastern and Western markets. To be sure, these activities are founded on purely economic bases; in the process, however, they are conducive to economic interdependence – tantamount to political influence.

In January 1966, Iran and the Soviet Union concluded an agreement according to which Iran would provide Russia with natural gas for 15 years beginning in 1970 at a rate of 10 billion meters every year. The agreement included the construction of a 750-mile pipeline for the transportation of gas from the oilfields in the Gulf to the city of Astara inside the Soviet border on the Caspian Sea. The $450 million Iranian Gas Trunkline (IGAT) was completed in September 1970, and the flow of gas began. In return, the USSR provided Iran with a $260-million credit to finance the construction of a steel mill at Isfahan and a machine-tool plant at Arka.[25]

Although the development of the Iranian natural gas is defended on its economic value, the significance of this agreement, nevertheless, lies in its starting with this particular project. The project was carefully targeted to present the Iranians with the extra cash from what the West used to put into flames for years. To convert into economic value what had previously been wasted entails a state of appreciation, conducive to harmonious relations and cooperation. Certainly the project did not escape the notice of the states around the Gulf. They, too, want to convert waste into economic value.[26]

A second 15-year agreement was signed between Iran and the Soviet Union in October 1970. It called for an increase in the exportation of Iranian natural gas, the construction of a second pipeline, and the expansion of the Isfahan steel plant. The agreement also provided for joint oil and gas exploration and joint petrochemical ventures. It is also reported that the Soviet Union provided Iran with a loan of $44 million to encourage the importation of Russian capital goods.[27]

The building of a 42-inch pipeline from the Iranian town of Bushir on the Persian Gulf to Astara to supply the Soviets with 1.4 million barrels of crude oil a day was the result of a third agreement between Russia and Iran.[28] It should be pointed out that Western production from Iran averaged four million barrels a day in 1970. The new agreement with the Soviet Union represented over 37 percent of the oil produced by Western companies in Iran.

In the meantime, the base of economic cooperation between the two countries was further accelerated. In 1970, a trade agreement between Iran and the Soviet Union was signed under which Iran would export goods equivalent to the amount of $720 million over a five-year period. In return, Iran would receive Russian goods valued at $318 million. The difference represented repayment by Iran of loans used for industrial projects. In 1964, Iran accounted for only seven percent of the Russian export to the region. By 1969, Iran took 21 percent, thus becoming the second largest Russian market in the area.[29]

An agreement was signed in 1967 for the purchase of £40,000,000 worth of Russian munitions. This was the first time the Russians concluded an arms agreement with a member of the Western bloc.[30]

Like Egypt among the Arabs, Iran is the most probable candidate to shape the politics of the Gulf. Among the Gulf states, Iran is the most heavily populated. The country enjoys a degree of political stability and economic diversity. Like Egypt, Iran is now the scene of a renewed power struggle for polarization; for 'it is in her relations with Iran [that] the Soviet Union's actual interest in frontier security and her potential interest in oil resources most clearly converge, to be joined there by the influence of history of active Russian involvement which stretches back beyond the Bolshevik Revolution.'[31]

Soviet inroads in the Persian Gulf did not, however, stop at Iran; they continued to Iraq. With Iraq, the Soviet Union concluded an oil agreement which, according to petroleum specialist Ruth Sheldon Knowles, amounted to 'an economic coup d'etat.'[32] The agreement was hailed by Soviet commentators as an instance in which Arab oil-producing countries would be able to evolve 'an oil policy which accords with their national interests.'[33]

Like Iran's natural gas agreement, the Iraqi agreement was carefully targeted to provide another shock in the control of oil by the West. The agreement, signed December 1967, called for the provision of £28 million to help Iraq National Oil Company (INOC) develop the North Rumaila oil field, originally discovered by the Iraq Petroleum group.[34] Now that Western companies control the transportation facilities and marketing outlets, INOC would have a difficult time selling the oil. Hence came the Russian novelty. It amounted to an injection in the agreement of a guarantee for the transportation and marketing of the oil. This agreement with Iraq was a turning point in which 'Moscow,' according to Edward Hughes, 'stunned the oil world.'[35] The agreement with Iran helped the Iranians convert into economic value what the West used to put into waste. The one with Iraq helped the Iraqis break away from Western control. Both were at the expense of Western influence.

In June and July 1969, two other oil agreements were signed. Under these agreements the Soviet Union provided Iraq with $140 million worth of materials and technical aid in exchange.[36]

Among the Arab countries Iraq ranks the third in its economic relation with the Soviet Union.[37] It is also known that the bulk of Iraq's military hardware is supplied by the Soviet Union.

In October 1970, an exchange deal was signed between Kuwait and the Soviet Union. Under this deal Kuwait is to provide

Russia's customers in Asia with refined oil against Russia's supplies of the same material to Kuwait's customers in Europe. The agreement was hailed by the Kuwaitis as 'a first step in broader co-operation in oil.'[38]

Saudi Arabia is very cautious in her dealing with the Soviet Union. She nevertheless signed with Rumania an oil barter agreement.[39]

In South Arabia the Russians made clear headway in both Yemen and South Yemen People's Republic. The latter is the scene of an organized guerrilla movement whose aim is 'the liquidation of Western interests in the occupied Gulf.'[40] The Russians' interests in South Arabia are 'not dissimilar to those that first induced Britain to establish bases there.' The strategic location 'as a land bridge to East Africa and as a gateway to the Persian Gulf and the Indian Ocean' could hardly escape the notice of the Soviet Union and/or China for that matter.[41]

The Persian Gulf is now the scene of a renewed power struggle, the outcome of which is not clearly determined. T.B. Millar hints at the direction of events with a question: 'How likely would the Soviet Union be to seek to replace Britain in the Gulf, or to stimulate local nationalist activities against Western interests?' In direct reference to what happened over the Suez Canal Millar elaborates:

> The precedents in the area and elsewhere are not comforting and no other state would be powerful enough, near enough and interested enough.[42]

The situation over the Suez Canal is, in many ways, relevant to the discussion of the significance of oil, and, as such, a parallel between the two is in order.

Similar to their treatment of the importance of oil, Western observers like to belittle the importance of 'the world's most important manmade waterway'[43] by injecting into the discussion the benefits the Russians would have gained had the Canal been opened. According to Millar the Canal 'is not, of course, the most important waterway for the Soviet Union, currently or potentially.'

He asserts that 'Before its closure in 1967, only a small fraction of Soviet trade went through the Suez Canal; this would not significantly increase if the Canal were reopened in the future.'[44]

Against this background of limited Soviet use of the Suez Canal, ten Western European foreign ministers from the European Common Market, including Britain, Ireland, Norway and Denmark, gathered together in May 1971, to devise a plan that called for the immediate withdrawal of Israeli troops from the occupied Arab lands, the administration of the holy city of Jerusalem by an international body, and the establishment of demilitarized zones on both sides of the 1967 Arab–Israeli borders. The motivating force behind the ministers' gathering was 'their concern at the continued closure of the Suez Canal which forces oil for Europe to be carried an extra 12,000 miles around Africa, adding greatly to its cost.'[45] To get a sense of the cost burden, Wilson M. Laird, the Director of the Office of Oil and Gas in the US Interior Department, put the freight rates from the Persian Gulf to the US East Coast at $3.30 per barrel of crude as compared to only 52 cents before the closure of the Suez Canal in June 1967.[46]

Between these Soviet limited uses and European extensive ones, the Suez Canal remains closed and, most importantly for our discussion, the Russians have to give their permission for its rectification. For what reasons, then, are the Russians there? Millar notes the obvious similarity between their position on the Suez Canal and their recent drive in the direction of the Persian Gulf, saying:

> From the Soviet view point, it [the Suez Canal] must be seen primarily for its strategic value, and this raises the whole question of Soviet policies in the Persian or Arabian Gulf and elsewhere in the Indian Ocean.[47]

The Russians are now in the Mediterranean and on the Suez Canal. They are at Aden and on the island of Mauritius. Their recent drive towards the Persian Gulf is, according to Professor Halford L. Hoskins, none other than a renewal of an old objective put forward by the Soviet Secretary Vyacheslav Molotov to the

German Ambassador Count von Schulenberg on the eve of World War II. In a memorandum dated November 25, 1940, the Russians demanded a 'recognition of Soviet aspirations in the direction of the Persian Gulf.'[48]

This, as noted earlier, is the precise course of action followed by the British at the time of their imperial expansion, but with one very big difference: The British advanced their expansion by the force of arms. In so doing, they planted hatred and ploughed resistance. The Russians, on the other hand, advanced their expansion 'by invitation.'[49] They were called upon to liberate and protect the oppressed. The difference between the two approaches is, by and large, due to American blunders.

The concern over the Soviet drive towards the Persian Gulf is sometimes belittled by placing undue emphasis on Russia's domestic need for foreign oil. Figures are often provided to substantiate the claim. It has been said, for example, that even in 1980 Soviet need for foreign oil (estimated at 100 million tons or so) would be less than the 1968 oil importation of either Japan or Western Europe – put at 132 million tons and 253 tons, respectively.[50]

But this applies equally to the United States. The United States now controls 60 percent of the oil resources of the area, with almost none of these resources reaching the shores of the United States. Could it be said that the United States is not benefiting from the control of the oil resources of the Gulf?

Oil is vital to the continued growth and prosperity of Western Europe and Japan. The hegemony of the Persian Gulf as a source of supply is indisputable. According to one State Department official 'the Arab oil-producing nations will hold sway' at least for the next 10 to 15 years. He went on to say that, 'if there were a concerted boycott by the nations of the Persian Gulf and North Africa, it could cause a tremendous problem for the free world.'[51]

Obviously, these facts were known to the Soviet Union, and it is rather missing the point to refer to the Soviet domestic need for foreign oil as the motivating force behind the Soviet drive in the Persian Gulf. Apparently, this line of reasoning exasperated Waldo

H. Dubberstein of the Central Intelligence Agency. Before the Princeton University annual conference, Dubberstein declared that 'Russia is in the Middle East because that is the background against which Soviet extension into the Persian Gulf will occur.' And she is in the Persian Gulf because 'her vital interests are in Europe.' Dubberstein then wedded the two situations together:

> We cannot sit here and complacently say that the Russians have oil, that they are not going to move into Middle East oil, that they can't market it, that they can't do a lot of other things. The Russians don't need the oil but they know that control of Middle East oil brings with it certain important control elsewhere in the world.[52]

Western Europe and Japan will continue to be Russia's foremost targets in her three-dimensional policy. Indications are that the Soviets are determined to hold onto the portion of the oil market they gained in these countries. Oil is a hard currency earner. It is also an instrument of political control.

In 1967, the Russians exported 48 million tons of crude oil to Western Europe. This is about 21 percent of Western Europe's import from the Persian Gulf area.[53] Italy is the largest importer of Russian oil, followed by Germany who in 1969 agreed with the Russians to the importation of eight million tons of oil per year. Russia's supplies of oil and oil products reached Ireland, Spain, Belgium, Denmark, and Turkey. Japan, of course, stands out as the leading importer of Russian oil in the Far East. In 1966, Japan imported 100,000 barrels per day. Deliveries dropped to 80,000 barrels per day in 1967 due to the closure of the Suez Canal.[54]

The Russians began to embark upon the expansion of their distribution facilities, Nafta Ltd. in Britain. It was reported that Nafta planned to open 500 service stations all over Britain. The Russians own a marketing organization in Belgium and they were building substantial new tankage in Antwerp. They were also considering the purchase of refineries in some Western European countries. In 1969, an exchange deal was signed between the Soviet Union and West Germany in which Germany undertook the construction of two pipe manufacturing plants for the production of 100-inch diameter pipeline for the transportation of natural gas from Siberia in exchange for Russian oil and natural gas.[55]

Western Europe, of course, will not allow itself to become too dependent on Soviet oil supplies. The Russians are, nevertheless, striving hard to capture a larger and larger segment of the Western Europe energy market. In their approach, they are very cautious, though, on occasion, hints of their strategy appear. In commending the development of hydrocarbon reserves in Siberia, *Izvestia*, the official Soviet Government newspaper, stated that the development of these resources would 'make a significant contribution to the USSR's economy.' 'It will also,' said the paper, 'have a beneficial influence on our economic ties with other nations.'[56]

Siberia is overwhelmed with economic and technical difficulties. Thus with the exhaustion of the old Baku oilfields and with oil supply, in general, lagging behind demand, there is a limit to the Soviet domestic capacity. The sign of this limitation appeared in 1967 when, for the first time, the rate of Soviet oil export expansion was leveled out and was expected to decline.[57] Indeed, in 1970, the Soviets were forced to consider a limitation of their oil supply to the countries of Eastern Europe at 680,000 barrels a day (over half of these countries' oil requirement of 1.2 million barrels a day) so as to hold on to their portion of the growing oil market of Western Europe.[58]

In view of the difficulty in keeping supply even with the rising demand, and in view of the Soviet insistence to remain 'a permanent fixture in Western oil markets,'[59] a look across the border in the direction of the cheap oil resources of the Persian Gulf appeared to be the most tempting and the obvious next stop.

Eastern Europe is, of course, the third target in the Soviet oil and natural gas policy. These countries – Poland, East Germany, Czechoslovakia, Hungary, and Bulgaria – are heavily dependent on solid fuels as sources of their energy requirements. In 1966, solid fuels formed 88 percent of these countries' total energy. The rest was covered by oil, 9.5 percent; natural gas, two percent; and primary electricity, one-third percent.[60]

This energy composition is, of course, a sign of deficiency in modern fuels and, therefore, a loss in competitive advantage. Accordingly, the East European countries, in the interest of economic efficiency, were striving to increase the proportion of hydrocarbon fuel in their energy consumption.

The Eastern bloc as a whole, including Rumania and Albania, but excluding the Soviet Union, consumed around 35 million tons of oil and 25 billion cubic meters of natural gas in 1967. This is 21 percent of the total bloc consumption. In 1970, the bloc's oil consumption was 1.2 million barrels a day or 60 million tons per year. A substantial rise in the share of the hydrocarbon fuel is predicted in 1980. Oil consumption in 1980 is expected to reach 81 million tons per year with natural gas consumption rising to 100 billion cubic meters, both forms of energy making a combined share of 40 percent of the total energy consumption of Eastern European countries.[61] Another estimate put the oil requirements of these countries at 170 million tons per year in 1980.[62]

With the exception of Rumania, which in all probability will continue exporting its oil surplus to countries outside the bloc and which itself is the scene of barter agreements with the oil-producing countries of the Middle East to supplement its production, production of hydrocarbons in the rest of Eastern European countries is insignificant. Thus appears the need for oil importation from sources outside the bloc, and 'at this point,' says the *Economist*, 'politics step in.'[63]

In 1967, Russia provided Eastern Europe with 25 million tons of oil. This is two-and-a-half times the quantity delivered by the Soviets in 1961 and represented 55 percent of Eastern Europe's petroleum requirements. Of the 60 million tons said to have been consumed in 1970, the bloc, mainly Rumania, produced only one-third. The Soviet Union then provided the bloc with 680,000 tons daily (34 million tons per year) or about 56.6 percent of its oil requirements.[64]

It is becoming difficult for the Soviet Union to satisfy its oil energy requirements at home, and those of the East and West from domestic production. At times, therefore, the Soviets endorsed the beginning of barter agreements between the countries of Eastern

Europe and the oil-producing countries of the Middle East. Steps were taken by the Hungarian and Czechoslovak governments in early 1968 to conclude barter agreements with Iran in exchange for the latter's oil. After the Czechoslovak crisis of 1968, however, the Soviet Union reversed its position, clearly in the interest of tighter control.

Apparently, the Soviets had settled for control of over half the oil requirements of the countries of Eastern Europe as an effective proportion for control purposes. Thus, says the *Economist*, 'By providing the bulk of east Europe's oil needs, Russia is in a strong position to exert pressure and, after the whip-cracking over Czechoslovakia, it is a fair guess that it will maintain this relationship, even if it means importing more oil itself.'[65]

Soviet oil imports, however, are becoming mandatory, and no other sources are near enough, prolific enough, and cheap enough to equal the oil resources of the Persian Gulf.

The Western withdrawal from the Gulf area is now complete with the British departure. There exists, nevertheless, an infinitesimal American military presence stationed on the island of Bahrain. The Russians certainly have respect for this American presence. They, however, have developed the techniques and the justifications to outflank it. Until now, the Russians have proved successful, and Laqueur noted that 'the Russian drive to the south which began in the eighteenth century seems at last likely to achieve fulfillment.'[66]

Endnotes

[1] William D. Brewer, 'United States Interests in the Persian Gulf,' *The Princeton University Conference and Twentieth Annual Near East Conference on Middle East Focus: The Persian Gulf, October 24-25, 1968*, Cuyler Young (ed.) (Princeton: The Princeton University Conference, 1968), 177-178; 'US Investment in Arab World Tops $2000 M.,' *Middle East Economic Digest*, XIV, No. 34 (August 21, 1970), 984.

[2] Brewer, op. cit., 179; *The Middle East and North Africa, 1970-71* (London: Europa Publications Limited, 1970), 3; 'US Investment in Arab World Tops $2000 M.,' op. cit., 984.

[3] Brewer, op. cit., 178; 'US Investment in Arab World Tops $2000 M.,' op. cit., 984.

[4] *The Middle East and North Africa, 1970-71*, op. cit., 3; Brewer, op. cit., 178; 'US Investment in Arab World Tops $2000 M.,' op. cit., 984.

[5] Brewer, op. cit., 178; 'US Investment in Arab World Tops $2000 M.,' op. cit., 984.

[6] The Center for Strategic and International Studies, *The Gulf: Implications of British Withdrawal*, Special Report, Series No. 8 (Washington DC: Georgetown University, 1969), 6.

[7] John Allan May, 'British and US Strategy: Thwarting Soviet Influence,' *Christian Science Monitor* (November 17, 1970), 13; Russia Drives East of Suez,' *Newsweek* (January 18, 1971), 27.

[8] See Charles A. Heller, 'The Strait of Hormuz – Critical in Oil's Future,' *World Petroleum*, XL, No. 11 (October, 1969), 24-25.

[9] John Drysdale, 'South Asia Wants An Open Back Door,' *Christian Science Monitor* (November 20, 1970), 11; Paul Dold,

'African Reaction: Power Plays on Black, White Tensions,' *Christian Science Monitor* (November 18, 1970), 11.

[10] Walter Laqueur, 'Russia Enters the Middle East,' *Foreign Affairs*, XLVII, No. 2 (January, 1969), 291.

[11] Dold, op. cit., 11; May, op. cit., 13; George W. Ashworth, 'Russians in the Indian Ocean: Assessment from Washington,' *Christian Science Monitor* (November 17, 1970), 13.

[12] Laqueur, op. cit., 303.

[13] Quoted in May, op. cit., 13.

[14] Quoted in Maximilian Walsh, 'Australia Expands Defense Role to West,' *Christian Science Monitor* (November 20, 1970), 11; see also, Laqueur, op. cit., 307.

[15] Robert E. Hunter, *The Soviet Dilemma in the Middle East. Part II: Oil and the Persian Gulf*, Adelphi Papers, No. 60 (London: The Institute for Strategic Studies, 1969), 7; Walter J. Levy, 'Oil Power,' *Foreign Affairs*, XLIX, No. 4 (July, 1971), 664.

[16] Ashworth, op. cit., 13.

[17] Quoted in Edith Kermit Roosevelt, 'Oil, Arabs and Communism,' *America*, CXIX (September 21, 1968), 216.

[18] Robert E. Hunter, op. cit., 7.

[19] Paul Cohen, 'The Erosion of Surface Naval Power,' *Foreign Affairs*, XLIX, No. 2 (January, 1971), 331.

[20] Ibid., 332.

[21] Ibid.

[22] Ibid., 331.

[23] T.B. Millar, 'Soviet Policies South and East of Suez,' *Foreign Affairs*, XLIX, No. 1 (October, 1970), 70; Laqueur, op. cit., 296; 'Importance Growing As Supplier of Crude Oil,' *World Oil*, CLXXI, No. 3 (August 15, 1970), 166-171.

[24] See David Morison, 'Soviet Involvement in the Middle East: The New Strategy,' *Mizan*, XI, No. 5 (September-October, 1969), 263-264.

[25] Edward Hughes, 'The Russians Drill Deep in the Middle East,' *Fortune*, LXXVIII, No. 1 (July, 1968) 104; *The Middle East and North Africa, 1969-70* (London: Europa Publications Limited, 1969), 38 and 259; 'Growing Soviet Economic Stake in Middle East,' *Middle East Economic Digest*, XIV, No. 55 (August 28, 1970), 1009-1011.

[26] Robert E. Hunter, op. cit., 11.

[27] *Middle East Economic Digest*, XIV, No. 42 (October 16, 1970), 1212.

[28] *Middle East Economic Digest*, XIV, No. 2 (May 15, 1970), 577-578.

[29] *Middle East Economic Digest*, XIV, No. 33 (August 14, 1970), 958; 'Growing Soviet Economic Stake in Middle East,' *Middle Last Economic Digest*, XIV, No. 35 (August 28, 1970), 1009.

[30] *The Middle East and North Africa, 1969-1970*, op. cit., 256.

[31] Robert E. Hunter, op. cit., 1.

[32] Ruth Sheldon Knowles, 'A New Soviet Thrust,' *Mid East: A Middle East and North Africa Review*, IX (December, 1969), 5.

[33] Quoted in *Mizan*, XI, No. 3 (May-June, 1969), 184-185.

[34] 'North Rumaila Oil for Russia,' *Petroleum Press Service*, XXXVI, No. 8 (August, 1969), 284-286; *The Middle East and North Africa*, 1969-70, op. cit., 297; *Middle East Economic Digest*, XIV, No. 36 (September 4. 1970), 1044.

[35] Hughes, op. cit., 104; Robert E. Hunter, op. cit., 11-12; 'Russia and Arab Oil,' *Petroleum Press Service*, XXXV, No. 2 (February, 1968), 53.

[36] *Middle East Economic Digest* XIV, No. 24 (June 12, 1970), 704-705.

[37] Laqueur, op. cit., 303.

[38] *Middle East Economic Digest*, XIV, No. 43 (October 23, 1970), 1242.

[39] *Middle East Economic Digest*, XIV, No. 20 (May 15, 1970), 578.

[40] *Middle East Journal*, XXIV, No. 2 (Spring, 1970), 196-198.

[41] Laqueur, op. cit., 303.

[42] Millar, op. cit., 79.

[43] Ibid., 72.

[44] Ibid.

[45] See Richard H. Boyce, 'Mideast Peace Settlement Eyed,' *Rocky Mountain News* (May 21, 1971), 17.

[46] *Oil and Gas Journal*, LXVIII, (August 3, 1970), 48.

[47] Millar, op. cit., 72.

[48] Halford L. Hoskins, 'Changing of the Guard in the Middle East,' *Current History*, LII, No. 306 (February, 1967), 66.

[49] Laqueur, op. cit., 296; Millar, op. cit., 79.

[50] Robert E. Hunter, op. cit., 9; O.M. Smolansky, 'Moscow and the Persian Gulf: An Analysis of Soviet Ambitions and Potentials,' *The Princeton University Conference and Twentieth Annual Near East Conference on Middle East Focus: The Persian Gulf, October 24-25, 1968*, T. Cuyler Young (ed.), (Princeton: The Princeton University Conference, 1968), 154.

[51] Quoted in Monty Hoyt, 'Oil Output on Verge of Decline,' *Christian Science Monitor* (June 7, 1971), 1.

[52] Waldo H. Dubberstein, 'Comment,' *The Princeton University Conference and Twentieth Annual Near East Conference on Middle East Focus: The Persian Gulf, October 24-25, 1968*, T. Cuyler Young (ed.), (Princeton: The Princeton University Conference, 1968), 169-170.

[53] See Robert E. Hunter, op. cit., 3.

[54] Erhard Hagemann, 'Germany Weighs Soviet Crude Offer,' *World Petroleum*, XL, No. 8 (August 1, 1969), 36. 'Soviet Export Rise Halted,' *Petroleum Press Service*, XXXV, No. 6 (June, 1968), 205-206; Hughes, op. cit., 104.

[55] Huqhes, op. cit., 104; 'Soviet Export Rise Halted,' op. cit., 206; Hagemann, op. cit., 34.

[56] Quoted in Frank J. Gardner, 'Soviets Chortle Over Gas Riches, US Supply Pinch,' *Oil and Gas Journal*, LXVIII, No. 36 (September 7, 1970), 53.

[57] Hughes, op. cit., 104; 'Soviet Export Rise Halted,' op. cit., 205.

[58] 'Soviet Oil Strategy,' *Middle East Economic Digest*, XIV, No. 47 (November 20, 1970), 1352.

[59] Hughes, op. cit., 104.

[60] 'The Soviet Bloc is Hardening,' *Petroleum Press Service*, XXXV, No. 11 (November, 1968), 404.

[61] 'Soviet Oil in the 'Seventies,' *Petroleum Press Service*, XXXVII, No. 1 (January, 1970), 5; 'Soviet Oil Strategy,' op. cit., 1352.

[62] Christopher Tugendhat, *Oil: The Biggest Business* (New York: G.P. Putnam's Sons, 1968), 253.

[63] 'When Oil Flows East,' *The Economist*, CCXXXIV, No. 6594 (January 10, 1970), 51.

[64] 'The Soviet Bloc is Hardening,' op. cit., 405; 'Soviet Oil Strategy,' op. cit., 1352.

[65] 'When Oil Flows East,' *The Economist*, op. cit., 51.

[66] Laqueur, op. cit., 296.

7

Wresting The Control: The Organization of the Petroleum Exporting Countries

No sooner had the fallout of the Iranian crisis settled down than the oil-producing countries were stunned again by reductions in the posted oil prices. Without consulting the oil-producing countries, the international oil companies in February, 1959, reduced the posted prices of crude oil by an average of $0.18 per barrel.[1] Posted prices are 'transfer prices' used by a major international company among its affiliates and serve as a basis for determining the income of the producing stage upon which the revenues of the oil-producing countries are determined. The reduction, then, meant a substantial decrease in the revenues of the oil-producing countries, especially because 'a drop of one United States cent [per barrel] would mean $35 less in fiscal revenue, and to make that up, the area would have to increase production by 40 million barrels a year.'[2]

The oil-producing countries, of course, were angered at this measure, and at the first Arab Petroleum Congress, held in Cairo in April 1959, a resolution was adopted in which a recommendation was introduced calling for the companies to consult the oil-producing countries before they effect any changes in the posted prices.[3]

The international oil companies, however, disregarded this resolution, and in August 1960, they again reduced the posted prices by about $0.09 per barrel, bringing the prices down by about $0.27 per barrel over a period of 19 months. As a result of these two price cuts, the revenue per barrel of the oil-producing countries was reduced by $0.13, or 15 percent of the level at the beginning of 1959.[4]

The companies' behavior called for the immediate reaction of the oil-producing countries. On September 14, 1960, a conference, attended by representatives of the five principal oil-exporting countries – Iran, Iraq, Kuwait, Saudi Arabia, and Venezuela – met in Baghdad at the invitation of the Iraqi Government. The birth of the Organization of Petroleum Exporting Countries (OPEC) was the result of the first resolution adopted at this conference.[5] Subsequently, OPEC membership was extended to include Indonesia, Libya, Qatar, Abu Dhabi, Algeria, and Nigeria.

The Organization amassed the highest concentration of oil operations the world over. Its ten members were in 1970 responsible for 77 percent of the free world's oil reserves, 56 percent of its production and, more significantly, 90 percent of its crude oil exports.[6] This is an economic power with a decided impact on the business of oil.

The first resolution that established OPEC also declared the objectives of the Organization. From its inception, OPEC conceived of its function, as embedded in Paragraph 4 of its Resolution I.2, broadly to aim at 'the unification of petroleum policies for the Member Countries and the determination of the best means for safeguarding the interests of Member Countries individually and collectively.'[7] To achieve this aim, the objectives of the Organization were then divided into short- and long-term. An examination of the short-term objective indicates that it was designed to maximize member governments' revenues from the value of their production. The long-term objective aims at the achievement of effective control over the oil operations by member governments so as to gear the industry in the direction of the interest of the oil-producing countries. Significantly, therefore, the first item in the first resolution stated that 'Members can no longer remain indifferent to the attitude heretofore adopted by the oil companies in effecting price modifications.'[8]

To put these objectives into effect, however, something has to be done to contain the adverse effect of the surplus capacity instrument through the regulation of production or, as the concept subsequently came to be known, the management of supply among member countries. The conferees, therefore, took notice of the instrument and warned against its effectiveness in the fourth item in their first resolution:

> That is as a result of the application of any *unanimous* decision of this Conference any sanctions are employed, directly or indirectly, by any interested Company against one or more of the Member Countries, no other Member shall accept any offer of a beneficial treatment whether in the form of an increase in exports or an improvement in prices, which may be made to it by any such Company or Companies with the intention of discouraging the application of the unanimous decision reached by the Conference.[9] (Emphasis added.)

This 'unanimous' requirement is a serious drawback in the containment of the surplus capacity instrument. For it means that a country not a party to the decision could, if tempted by the oil companies, increase its production to make up for the shortage caused by the action of another member country in dispute with the oil companies. The implementation of this resolution requires self-restraint and solidarity among member governments of the kind that would induce a country to turn away customers and forego additional cash in support of another member country.

Nevertheless, with this shortcoming in mind and with an understanding of the limits of the possible, OPEC organized itself to present the concept of collective bargaining in negotiating disputed points with the international oil companies.

The sources of disputes among producing countries and the companies are many. They all, however, stem from the absence in the contracts of a provision to bring an automatic adjustment in view of changing circumstances.

The principal concession agreements in the Persian Gulf region were granted when most of the countries of the area were politically subordinated, directly or indirectly, to foreign powers, and these contracts were drawn up with no concern for the national interests of the countries involved.

These concessions were formulated for long durations. The life of these concessions varies from a minimum of 40 years to a maximum of 92 years. The consortium agreement in Iran runs for 40 years, until 1994 (allowing for three five-year extensions); the Arabian American Oil Company (Aramco) in Saudi Arabia, for 66 years until 1999; the Iraq Petroleum Company (IPC) in Iraq, for 75 years until 2000 and 2007 and 2013 for its associates Mosul Petroleum Company and Basra Petroleum Company, respectively; Qatar Petroleum Company in Qatar, for 75 years until 2010; Abu Dhabi Petroleum Company, for 75 years until 2014; and Kuwait Oil Company (KOC) in Kuwait, for 92 years until 2026.[10]

The duration of these concessions is overly generous. But, while these agreements were designed to ensure the concessionaire a reasonable time to produce, they were also drawn up in great detail to 'regulate nearly all aspects of the relationship between the government and the concessionaire company, from such crucial matters as the setting of tax rates for the life of the concession to relatively insignificant questions of detail.'[11]

In effect, therefore, these contracts reduced the action of a sovereign government in the oil-producing countries to that of a mere 'tax collector.' The legislative arms of the governments concerned were prevented from formulating laws applicable to the companies; their judicial bodies had no jurisdiction over disputes emanating from the application of the contracts, and their executive branches could not alter regulations applicable to the companies as circumstances changed.[12]

Thus, in the absence of a specific provision to allow for an automatic reassessment to reflect changing circumstances, and with the long lives of the contracts, coupled with the feeling that they were drafted in the absence of legitimate authorities representing the countries concerned, modifications in the outdated terms of these contracts were almost always accompanied by eruptions of one sort or another. Here, then, is the importance of OPEC as a necessary machinery for ironing out difficulties through negotiations.

Before the establishment of OPEC the oil-producing countries had always been in negotiation with the oil companies.

But, while the international oil companies were presented as a bloc by virtue of their joint ownership of concessionaire companies in the Middle East, the oil-producing countries were negotiating on an individual basis, always with the weapon of surplus capacity in the background. The emergence of OPEC, therefore, introduced a new element in the negotiation which had hitherto been lacking on the part of the oil-producing governments, i.e., the element of collective bargaining.[13]

Thus, from its position as a source of 90 percent of the world's oil exports, OPEC began negotiating its short- and long-term objectives with the major oil companies. The major oil companies, however, were determined not to recognize OPEC's position. They boycotted its meetings and insisted on negotiating with individual governments. Their attitude was to continue until 1971, ten years after OPEC's establishment, when they finally agreed to negotiate with OPEC as a collective entity.

Despite the companies' attitudes OPEC made it clear that every negotiation was influenced and backed by its collective membership. It is on this basis OPEC began negotiating its short- and long-term objectives.

The first objective for OPEC is, as stated earlier, the maximization of member governments' revenues from the value of their oil production. Stabilization of the oil prices had obviously been the first target in OPEC's list to achieve this objective. This was followed by expensing the royalties, the elimination of marketing allowance and of all unjustifiable payments made by the governments to the companies.

OPEC in general had been successful in limiting the companies' ability unilaterally to reduce the oil price, though it took the Organization more than 10 years before the prices could be restored to the pre-February 1959 level. In sharp contrast to the upsurge in the prices of manufactured goods, oil prices remained stable at their low level during the 1960s. It is worth mentioning at this point that this phenomenon of low oil prices was, however, hidden from the final consumer by the imposition of high taxes on oil products by the governments of the oil-consuming countries such as Britain and Germany.[14] Thus, while the trade journals[15] hurried to blame the oil-producing countries for the price hike of 1970, an examination indicates that on the basis of the 1967 average price of one barrel of oil sold in Western Europe, 47.5 percent of the price component was levied in taxes by the governments of the oil-consuming countries in Western Europe. Of the aggregate value of that barrel, only 7.9 percent is collected in revenues by governments of the oil-producing countries (see Table 14).

The second target in OPEC's list in its attempts to maximize its member governments' revenue from the value of their oil production was the treatment of royalties as an expense rather than as a credit against income tax payment.

Royalties, established at 12.5 percent of the posted price, are considered intrinsic values paid to a third party for the exploitation of an asset, such as petroleum, which is in the nature of being wasted by extraction. This is a universally recognized principle which was accepted by the major oil companies in their dealings with the governments of the United States, Venezuela, and Canada.

Before the conclusion of the 50-50 formula in 1951, payments to the governments of the oil-producing countries were made in the form of royalties levied on a unit production. After the imposition of the 50-50 tax measure, the companies disregarded the principle of royalties as being an intrinsic value of a wasted asset paid to a third party. The companies considered the third party to be the same entity – governments of the oil-producing countries – and, as such, the distinction is not admissible. This, of course, resulted in lumping together the amount of royalties and taxes due, to equal 50 percent of the companies' net profit. If we assume a posted price of $1.80 and a cost of production of $0.20 for a barrel of oil the companies would be paying to the governments in taxes an amount of 80 cents under the 50-50 tax arrangement. The companies, however, were paying only 57.5 cents in taxes. They were computing the royalties at 22.5 cents (12.5 percent of $1.80) which was then subtracted from the 80-cents tax to bring it to 57.5 cents. Accordingly, the companies were 'either paying taxes at full rate of 50 percent, but paying no royalty; or they were paying royalty but their income tax amounts to about 40% of income only.'[16]

TABLE 14

Price Component of One Barrel of Oil (Based on the 1967 Average Price of One Barrel of Oil Sold in Western Europe)

	Absolute Value	Percent of Total
Cost of production	$0.285	2.7
Cost of refining	0.350	3.3
Tanker freight	0.080	6.3
Storage, handling, distribution and dealer's margin	2.790	26.0
Oil company's net profits	0.681	6.3
Indirect and turnover oil taxes in consuming countries	5.100	47.5
Revenue of producing countries	0.853	7.9
	$10.739	100.0

Source: OPEC, OPEC Bulletin *(September–October, 1969), 1.*

Thus at the fourth conference held in Geneva in April and July 1962, OPEC resolved (Resolution IV.33 of June 8, 1962) 'that companies enjoying in Member Countries the right of extracting petroleum which is a wasting asset should, in conformity with the principle recognized and the practice observed generally in the world, compensate the countries for the intrinsic value of such petroleum apart from their obligations falling under the heading of income tax.'[17]

At a posted price of $1.80 the financial effect of expensing the royalties would yield the governments of the oil-producing countries an extra 11 cents per barrel in revenue. There ensued, therefore, laborious negotiations between OPEC representatives and the oil companies. Negotiations reached a deadlock on a number of occasions when a last-minute offer was made by the major companies while OPEC member countries' representatives were attending their sixth conference in July 6-14, 1964. The companies finally agreed on a phase-out system for expensing the royalties. According to this system, the companies recognized the principle of expensing the royalties but attached to it a system of discounts from the posted price of 8.5 percent in 1964, 7.5 percent in 1965, and 6.5 percent in 1966. Further reduction in the discount after 1966 would be determined by market conditions.[18] Of the 12.5 percent royalties required by the governments to be expensed, the companies, in effect, agreed to phase out only four, five, and six percent in 1964, 1965 and 1966, respectively. In 1968, OPEC negotiated a schedule under which the discounts allowed to the companies were to be completely phased out by 1975.

With the objective of expensing the royalty partially achieved, OPEC moved to its third target in maximizing its members' revenues, i.e., the elimination of marketing allowances. Against the posted prices, the companies used to charge the producing stage with marketing expenses incurred on downstream operations. The companies, therefore, were transferring part of the sales value to the parent companies in the form of expenses to their subsidiaries. This deduction, of course, had the effect of reducing the tax base of the producing countries. OPEC saw no justification in penalizing the production stage by expenses incurred by successive stages and in 1962 Resolution IV.34 was passed, calling for the elimination of marketing allowances. In the end, this objective was successfully achieved.

The objectives outlined above were of immediate concern and direct benefit to all member countries. Therefore, OPEC had no difficulty in courting the support of all its members in achieving these objectives. However, the potential importance of OPEC lies in its ability to step, directly or indirectly through its member governments, into the management of the oil supply.

If OPEC is to stay active and productive so as to safeguard its member countries' interests, a joint production program as a means to curtail the effectiveness of the surplus capacity instrument is obviously needed. Member countries had, in their first resolution, warned against the possibility of being played off one against the other. Thus, in their attempts to institutionalize the management of supply, member countries authorized, in December 1964, the setting up of an OPEC Economic Commission to study the problems presented by the world supply situation. After two years of study, the Commission concluded that a joint production program, agreed upon by the producing countries, appeared to be the only cure against the surplus capacity. Therefore, a 'transitory' production program prepared by the Economic Commission was incorporated in Resolution IX.16 adopted in the OPEC conference held in Tripoli in July 1965. The plan called for 'rational increases in production from the OPEC area to meet estimated increases in world demand.'[19] Commenting on this resolution, *The Economist* said:

> It can be argued that this control of offtake is the one prerogative essential to management of an international oil company. But if it is governments, working to an agreed OPEC programme, who are to settle how much oil is taken from each country each year, then ultimately this power would have shifted to the governments.[20]

Programming production would entail the development of a situation in which a producing country would refuse the temptation of extra production increases and, consequently, extra foreign exchange necessary for economic development. The program was supported by Venezuela, which was more interested in higher prices than in expanding production. It was opposed by Iran, which was in the midst of her fourth economic plan, and by Libya, who had just entered the oil business. Both countries were interested in production expansion. The program therefore did not get off the ground, and at the OPEC conference of November 1967, the conference 'reaffirmed its conviction that a program is an effective instrument for the pursuit of the Organization's objectives, and instructed the Economic Commission to undertake a comprehensive study with a view to perfecting an economically practicable system.'[21] It is obvious from the wording of this statement that the program was shelved, at least temporarily.

But, while member countries could not reach a voluntary agreement on limiting their production to assigned shares to individual countries, there developed an upsurge in world oil demands, coupled with the closure of the Suez Canal and a shortage in tankers, and intensified by the Libyan oil cutbacks and the rupture of the TAP Line which in the end led to what petroleum specialist, Walter J. Levy, called the 'hurricane' of change.[22]

The base for this change was laid down in OPEC Resolution XVI.90 adopted at the sixteenth conference held in Vienna on June 24-25, 1968. This resolution is an important statement of objectives and policies. It sets the tone of OPEC future behavior and tells the thinking of member countries. The resolution is so comprehensive that its inclusion in full is presented below:

Sixteenth Conference
Vienna, June 24-25, 1968
Resolution XVI.90[23]

DECLARATORY STATEMENT
OF PETROLEUM POLICY IN MEMBER COUNTRIES
THE CONFERENCE
recalling Paragraph 4 of its Resolution 1.2; recognizing that hydrocarbon resources in Member Countries are one of the principal sources of their revenues and foreign exchange earnings and therefore constitute the main basis for their economic development;

bearing in mind that hydrocarbon resources are limited and exhaustible, and that their proper exploitation determines the conditions of the economic development of Member Countries, both at present and in the future;

bearing in mind also that the inalienable right of all countries to exercise permanent sovereignty over their natural resources in the interest of their national development is a universally recognized principle of public law and has been repeatedly reaffirmed by the General Assembly of the United Nations, most notably in its Resolution 2158 of November 25, 1966;

considering also that in order to ensure the exercise of permanent sovereignty over hydrocarbon resources, it is essential that their exploitation should be aimed at securing the greatest possible benefit for Member Countries;

considering further that this aim can better be achieved if Member Countries are in a position to undertake themselves directly the exploitation of their hydrocarbon resources, so that they may exercise their freedom of choice in the utilization of hydrocarbon resources under the most favorable conditions;

taking into account the fact that foreign capital, whether public or private forthcoming at the request of the Member Countries, can play an important role, inasmuch as it supplements the efforts undertaken by them in the exploitation of their hydrocarbon resources, provided that there is government supervision of the activity of foreign capital to ensure that it is used in the interest of national development and that returns earned by it do not exceed reasonable levels;

bearing in mind that the principal aim of the Organization, as set out in Article 2 of its Statute, is the coordination and unification of the petroleum policies of Member Countries and the determination of the best means for safeguarding their interests, individually and collectively;

recommends that the following principles shall serve as basis for petroleum policy in Member Countries.

MODE OF DEVELOPMENT

1. Member Governments shall endeavour, as far as feasible, to explore for and develop their hydrocarbon resources directly. The capital, specialists and the promotion of marketing outlets required for such direct development may be complemented when necessary from alternate sources on a commercial basis.

2. However, when a Member Government is not capable of developing its hydrocarbon resources directly, it may enter into contracts of various types, to be defined in its legislation but subject to the present principles, with outside operators for a reasonable remuneration, taking into account the degree of risk involved. Under such an arrangement, the Government shall seek to retain the greatest measure possible of participation in and control over all aspects of operations.

3. In any event, the terms and conditions of such contracts shall be open to revision at predetermined intervals, as justified by changing circumstances. Such changing circumstances should call for the revision of existing concession agreements.

PARTICIPATION

Where provision for Governmental participation in the ownership of the concession-holding company under any of the present petroleum contracts has not been made, the Government may acquire a reasonable participation, on the grounds of the principle of changing circumstances.

If such provision has actually been made but avoided by the operators concerned, the rate provided for shall serve as a minimum basis for the participation to be acquired.

RELINQUISHMENT

A schedule of progressive and more accelerated relinquishment of acreage of present contract areas shall be introduced. In any event, the Government shall participate in choosing the acreage to be relinquished, including those cases where relinquishment is already provided for but left to the discretion of the operator.

POSTED PRICES OR TAX REFERENCE PRICES

All contracts shall require that the assessment of the operator's income, and its taxes or any other payments to the State, be based on a posted or tax reference price for the hydrocarbons produced under the contract. Such price shall be determined by the Government and shall move in such a manner as to prevent any deterioration in its relationship to the prices of manufactured goods traded internationally. However, such price shall be consistent, subject to differences in gravity, quality and geographic location, with the levels of posted or tax reference prices generally prevailing for hydrocarbons in other OPEC Countries and accepted by them as a basis for tax payments.

LIMITED GUARANTEE OF FISCAL STABILITY

The Government may, at its discretion, give a guarantee of fiscal stability to operators for a reasonable period of time.

RENEGOTIATION CLAUSE

1. Notwithstanding any guarantee of fiscal stability that may have been granted to the operator, the operator shall not have the right to obtain excessively high net earnings after taxes. The financial provisions of contracts which actually result in such excessively high net earnings shall be open to renegotiation.

2. In deciding whether to initiate such renegotiation, the Government shall take due account of the degree of financial risk undertaken by the operator and the general level of net

179

earnings elsewhere in industry where similar circumstances prevail.

3. In the event the operator declines to negotiate, or that the negotiations do not result in any agreement within a reasonable period of time, the Government shall make its own estimate of the amount by which the operator's net earnings after taxes are excessive, and such amount shall then be paid by the operator to the Government.

4. In the present context, 'excessively high net earnings' means net profits after taxes which are significantly in excess, during any twelve-month period, of the level of net earnings the reasonable expectation of which would have been sufficient to induce the operator to take the entrepreneurial risks necessary.

5. In evaluating the 'excessively high net earnings' of the new operators, consideration should be given to their overall competitive position vis-a-vis the established operators.

ACCOUNTS AND INFORMATION
The operator shall be required to keep within the country clear and accurate accounts and records of his operations, which shall at all times be available to Government auditors, upon request.

Such accounts shall be kept in accordance with the Government's written instructions, which shall conform to commonly accepted principles of accounting, and which shall be applicable generally to all operators within its territory.

The operator shall promptly make available, in a meaningful form, such information related to its operations as the Government may reasonably require for the discharge of its functions.

CONSERVATION
Operators shall be required to conduct their operations in accordance with the best conservation practices, bearing in mind the long-term interests of the country. To this end, the Government shall draw up written instructions detailing the conservation rules to be followed generally by all contractors within its territory.

SETTLEMENT OF DISPUTES
Except as otherwise provided for in the legal system of a Member Country, all disputes arising between the Government and operators shall fall exclusively within the jurisdiction of the competent national courts or the specialized regional courts, as and when established.

OTHER MATTERS
In addition to the foregoing principles, Member Governments shall adopt on all other matters essential to a comprehensive and rational hydrocarbons policy, rules including no less than the best of current practices with respect to the registration and incorporation of operators; assignment and transfer of rights; work obligations; the employment of nationals; training programs; royalty rates; the imposition of taxes generally in force in the country; property of the operator upon expiry of the contract; and other such matters.

DEFINITION
For the purposes of the present Resolution, the term 'operator' shall mean any person entering into a contract of any kind with a member Government or its designated agency including the concessions and contracts currently in effect, providing for the exploration for and/or development of any part of the hydrocarbon resources of the country concerned.

This is a resolution with a sweeping impact on the changing relationship between the major oil companies and the oil-producing countries. One thing is clearly glaring in it: the pendulum of oil control is swinging in the direction of the producing countries. Basing their case on their first resolution of determining the best means for safeguarding their interests and on the principle expressed in the United Nations resolution of the 'inalienable right of all countries to exercise permanent sovereignty over their natural resources in the interest of their national development,' the oil-producing countries put forth the doctrine of 'changing circumstances' to effect their gradual control of the oil business. Specifically, the resolution called for:

1. the maximization of member governments' revenues through the right of the governments in determining the posted prices and what amounts to 'reasonable' earnings;
2. the control of oil through full ownership or participation in existing concessions;
3. the disputes arising from the application of these principles should fall under the jurisdiction of national courts.

The process, in motion for more than a decade, is picking up momentum now. Walter Levy, looking back to the days when he visited Iran as consultant to Averill Harriman, President Truman's envoy during the Iranian crisis, has this observation:

The winds of change for the oil industry that have been stirring throughout the decades since 1950 have now risen to hurricane proportions. The aim of the major oil-producing countries in this vortex is clearly to maximize their governments' 'take' out of the value of their oil production and obtain increasing control over oil operations. To achieve this, these countries – already formally joined in the Organization of the Petroleum Exporting Countries (OPEC) since 1960 – have now effectively combined to wield the economic and political power of an oil monopoly.[24]

The wheel was set in motion when a series of resolutions was adopted in OPEC's twentieth conference held in Algiers on June 24-26, 1970. In Resolution XX.112 member countries reaffirmed their faith in production programming. Resolution XX.113 called for 'full integration of the petroleum industry in the national economy of member countries.' Obviously this resolution aims at changing the status of the producing countries from producers of raw materials to being fully involved in the oil industry. The resolution called for repatriation in member countries of an adequate portion of the sale proceeds generated from exploitation of their hydrocarbons. Thirdly, Resolution XX.114 was adopted to reaffirm OPEC's total solidarity with Algeria in her negotiations with the French companies over price increases. Algeria had then taken a unilateral action in raising the price by 77 cents to $2.85.[25]

The biggest boost in implementing OPEC's resolutions, however, came as a result of the Libyan government's action in forcing a curtailment of the oil supply, thus achieving the same result of production programming sought by OPEC since its establishment. Libya lies west of Suez. With the closure of the Suez Canal in 1967, tanker rates soared, and Europe became dependent on Libyan oil for 30 percent of its supply. The Libyan bargaining power was further improved by the rupture of the Trans-Arabian Pipeline (TAP Line) in the Syrian desert on May 3, 1970. The pipeline was then responsible for pumping 450,000 barrels of oil per day from Saudi Arabia through Jordan and Syria to Lebanon on the Mediterranean. The TAP Line is owned by four American companies, owners of Aramco, and therefore the Syrian government refused its repair. The pipe remained broken for almost nine months. Additional enhancement of the Libyan position was provided by reasons of industry structuring. In Libya there are 23 oil companies, of which 17 are American, and only five of these are majors. The implication of this organization pattern is demonstrated in the lack of alternative sources of supply among the great majority of the companies operating in Libya.

It is within this context negotiation of a price between the Libyan government and the companies began, when the most that could be offered by the oil companies in increases was only between six and ten cents, and over a period of time. It was also reported that after the revolution the companies stepped up their production but reduced their drilling activities to half. In May 1970, the number of rigs in action was put at 26, half as many as the rigs that were in operation in the fall of 1969.[26]

This was an instance in which the Libyan government ordered, in the name of conservation, selected oil companies to cut back their oil productions. In May, 1970, Occidental Petroleum Corporation, an American concern, was ordered to cut its daily production from 800,000 to 500,000 barrels. Occidental was followed by Amoseas, owned by Standard Oil of California and Texaco, which was ordered in June to cut production by 100,000 barrels per day. Oasis, the largest oil producer, owned by Continental Oil, Marathon, Amerada (Americans), and Shell (Anglo-Dutch), was the third in the list to be ordered to cut its production by 150,000 barrels a day down to 895,000. Then, for the second time, Occidental was ordered to cut its production by a further 60,000 barrels a day.[27]

Amid these events, and while the Libyans were insisting that the cutbacks were necessitated by reasons of conservation policy, it was reported that Esso Standard and Occidental made an offer of higher prices which the Libyans considered to be way below their expectation. Subsequently Esso, the subsidiary of Standard Oil of New Jersey, too was ordered to cut its production by 110,000 barrels a day. Esso was followed by Mobil, ordered to cut production by 40,000 barrels per day.[28]

By August 1970, the Libyan production was down to 2,840,000 barrels per day from a previous average of 3,600,000. Thus, between the Suez Canal closure, the idleness of the TAP Line, and the Libyan cutbacks, the international oil supply was short by over one million barrels a day.[29] And that was enough to bring the companies to agree on the proposed increases in prices.

On September 4, 1970, Occidental Petroleum Corporation announced its acceptance of a rise in the posted price by 30 cents a barrel to $2.53. The agreement also provided for an increase of two cents a barrel every year until 1975. It too marked the end of the 20-year-old 50-50 profit-sharing formula. Occidental was followed by Esso, British Petroleum, Shell, and the rest of the oil companies operating in Libya. It was estimated that the agreement gave Libya an increase of $220 million a year in taxes and royalties.[30]

As expected, the Libyan agreement set off a chain of reactions in Iraq, Algeria, Caracas, Tehran, and again in Tripoli. On September 28, it was announced that Iraq succeeded in increasing the posted price of 20 cents per barrel for oil shipped from the Mediterranean terminal.[31]

Algeria, already considering its July offer of $2.85 (then not yet accepted by the French) out of tune with what the Libyans had secured, announced on October 30, 1970, that she raised the posted price to $3.24 a barrel. Oil talks between the French and the Algerians dragged on for 17 months before they broke off in February, 1971. At about that time (February 14, 1971) the Tehran agreement, which gave the producers of the Gulf area an increase of 35 cents per barrel, was concluded. A succession of moves then taken by the Algerians led, on February 24, to the decision to nationalize 51 percent of Compagnie Francaise des Petrole (CFP) and ELF-ERAP. The two French companies were offered $100 million in compensation; but not, however, before the outstanding debts due Algeria by the French companies were settled. Also, the Algerians raised the price of a barrel of oil to $3.60.[32]

The French were angered by these Algerian measures, and threatened to take the case to the International Court of Justice. They moved for a worldwide embargo on Algerian oil by warning foreign buyers. It was reported that the French had asked the United States to stop El Paso Natural Gas Company from buying huge quantities of Algerian natural gas since that gas could be from the fields of the former French companies.[33]

In the end, agreement between the two parties was concluded in accordance with the terms of the Algerian nationalization and the new posted prices. The French realized that

Algeria had its national oil company, Sonatrach, already in possession of its own technicians and expertise capable of running the oil business. They also saw the Americans proceeding with the conclusion of their natural gas deal with the American Export-Import Bank extending to Sonatrach loans, guarantees, and credits up to $299 million for the project.[34] The French blockade had crumbled before it started.

With the Libyan agreement concluded successfully, OPEC was emboldened to pass its Resolution No. XXI.120 at its twenty-first conference held in Caracas in December, 1970. Among other things, the resolution called for:

1.   The end of the 50-50 formula in favor of higher taxes.
2.   Increases in posted prices to reflect elimination of existing disparities between member countries on the basis of the highest price attained by any member countries.[35]

In accordance with the term of this resolution, representatives of Iran, Iraq, and Saudi Arabia met in Tehran on January 11, 1971, and called on the companies to negotiate on the basis of the Caracas resolution. From the beginning, it was obvious that the Shah of Iran took a personal interest in the negotiations. In one of his news conferences he was reported as saying that unless the companies and producer nations reached a new price agreement, the ten members of the Organization of Petroleum Exporting Countries would join in countermeasures that 'might be anything, including a stoppage of oil.'[36]

Similarly, Washington was directly concerned over the progress of the negotiations. The situation was described 'as potentially more serious than any yet faced by Western Europe and US since imported oil attained such a major role in these countries' energy supply.' Soon, President Nixon sent the Undersecretary of State, John N. Irwin II, to express the President's concern to the officials in Saudi Arabia, Iran and Kuwait. The United States then extended its invitation to representatives of the Organization for

Economic Cooperation and Development (OECD) and Japan to convene in Washington to meet the oil-producing countries with a united front. In addition, the United States gave antitrust clearance to her companies to negotiate en bloc with the oil-producing countries.[37] The oil companies gathered together in New York to draw up what was then called in Europe the 'cartel declaration.' The New York declaration, then signed by Jersey Standard, California Standard, Gulf, Mobil, Texaco, Shell, British Petroleum, and Compagnie Francaise des Petrole, plus seven other American independent companies, was delivered to OPEC representatives on January 16, 1971, 'with concurrence of the governments of the United States, Britain, France, and Japan.' The declaration watered down a call for OPEC to negotiate on behalf of the oil-producing countries – a development which OPEC had tried, but failed, to achieve for over ten years. The declaration also called for a five-year stability period in the oil business.[38]

A month of hard bargaining had to elapse before the two parties could sign the famous Tehran agreement on February 14, 1971. This settlement called for:

1. stabilization of the tax rate at 55 percent of the net income of the companies;
2. an immediate increase in posted prices of 35 cents per barrel for the oil shipped from the Gulf terminals;
3. an increase in the posted prices of five cents per barrel on June 1, 1971, and on the first of each of the years 1973-75, to reflect increasing demand for crude oil;
4. an increase of 2.5 percent in posted prices on June 1, 1971, and on the first of each year of the years 1973-75 to reflect adjustment for inflation;
5. the establishment of a new system of gravity differentials. For crude oils between 40 degrees and 30 degrees API Gravity each, the present posted price will be increased by one-half cent per barrel for each full degree;

6.   the agreement is to hold for five years through
      1975.[39]

Financially, the Tehran agreement yielded the Gulf states an estimated additional revenue of over $1,200 million in 1971, rising to about $3,000 million in 1975.[40] More important, however, was the chain of reactions set in motion by the conclusion of the Tehran agreement.

In April 1971, Libya had another round of talks with the oil companies in which Libya secured an increase in the posted price of 90 cents to $3.45 per barrel, retroactive to March 20, 1971. She also matched the progressive increases attained by the Gulf states. The tax rate was similarly fixed at 55 percent. At a production rate of three million barrels a day, the agreement would yield Libya an average increase of $708 million per year.[41]

On April 13, Algeria raised its posted price to $3.60, or 73 percent above the level of $2.08 attained in 1965, retroactive to March 20.[42]

Negotiations then moved to Iraq and Saudi Arabia to determine new posted prices for these two countries' crude oil pumped to ports on the Eastern Mediterranean. Iraq settled for an increase of 80 cents for a price of $3.21 per barrel retroactive to March 20. Similarly, Saudi Arabia received the same treatment. Generally, both countries received the same financial progression secured in the Tehran agreement – tax rate fixed at 55 percent, annual increase of five cents per barrel, and an annual increase of 2.5 percent for inflation.[43]

Thus the ten-year OPEC objective of restoring posted prices to their pre-1959 level had at last been attained. The power of OPEC, as we know, emanates from the development among its member countries of a feeling that their common interest is best served by their harmony. On the other hand, OPEC can hope to have its objectives truly attained by virtue of its members' greater understanding of the nature of the business they are in. Only then will they hold back from criticizing each other's interests. OPEC success, therefore, would be boosted by its members' direct involvement in the business of oil. Today the oil-producing countries are averse to being relegated to the status of an absentee landlord whose greatest interest is the collection of taxes. They want to have a say in the management of oil. They, therefore, developed their own national oil companies; they looked for direct producing-government to consuming-government relations; and they use the doctrine of direct exploitation or participation as the vehicle for achieving effective control.

As of now most of the oil-producing countries have their national oil companies equipped with a corps of technicians and the expertise necessary to provide them with the required 'know-how.' The National Iranian Oil Company was organized in 1951. It was followed by Kuwait National Petroleum Company, organized in 1960; Iraq National Oil Company, organized in 1961; and Saudi Arabian General Petroleum and Mineral Organization, organized in 1962. The Libyan National Oil Company was organized in 1968. These companies were given the responsibilities of exploring for and developing oil fields in partnership or in contractual relations with foreign companies. They were assigned the functions of refining, transporting and selling the oil in their home market. Eventually, the national oil companies would be the instrument through which the oil-producing countries would acquire effective control over the industry. It is in this context one must consider the statement of the Shah of Iran that 'The production of oil is not difficult. We are doing this now. Other countries can do this as soon as they have technicians. They can drill oil wells and fill the tankers. Why must the foreign companies do this?' The Shah went on to say that it is not important who buys Iranian oil; but rather that within the next few years purchases must be made directly from the producing countries.[44]

The process of achieving control over the industry is accelerated by direct government-to-government dealings. There had been a number of direct barter agreements between the oil-producing countries and almost all of the Eastern European countries. Almost every Western European country has its own oil company which is in direct contact with the oil-producing countries. Professor M.A. Adelman of the Massachusetts Institute of Technology watches this mushrooming of Western European oil companies and attributes it to the belief among the Europeans that their security problem emanates from their being placed at the mercy of huge Anglo-Saxon companies. Professor Adelman calls this European belief a myth. It stems from 'prejudice against big business, and against foreigners.' Adelman continues: 'Xenophobia is not only wrong but, like most prejudices, expensive.'[45]

While Mr. Adelman had put his finger on the cause of the European fear as being dominated by the major oil companies, his interpretation has not occurred, to the Europeans, or at least, are not yet at issue. Let us first follow Adelman's diagnosis of the European security problem.

Resting his analysis on the assumption that the Arabs had preconceived motives against the Europeans, Adelman advises the Europeans to be prepared for a total secession of oil by the oil-producing countries for a limited period – say six months. This, for Adelman, is both the problem and its remedy. How would the Europeans prepare themselves? For economic reasons, Adelman tells them to phase out their coal industry and send close to a million mineworkers off their jobs. He further tells them not to diversify their oil supply sources, for that is 'foolish.' Then he recommends a measure which he himself had said the Europeans had already done but for a shorter period, i.e., stockpiling the oil. Professor Adelman wants the Europeans to build oceanside terminals large enough to stockpile oil for a period of six months. He knows, of course, that the cost of building these facilities is staggering. He, therefore, wants the 'storage and oil to fill it [to] be provided at government expense.' 'But,' he goes on to advise the Europeans, 'for the sake of economy, private enterprise should manage the facilities and commingle [government] oil freely with theirs.'[46]

Petroleum Press Service, a British Journal, probably understanding that Adelman's report is a masterpiece in what it sets itself to warn against – prejudice – had this to say in introducing the report to its European readers: 'It should be considered *dispassionately* by all who are concerned with energy policy.'[47] (Emphasis added.)

The issue with Adelman and, for that matter, with petroleum specialist Walter J. Levy, who maintained that government-to-government oil dealing approach is 'disastrous' because it turns every economic problem into a political one,[48] stems from their underlying basic premises. Adelman chose to start his analysis from the point that the Arabs, or the Moslem countries, as some of the trade journals like to call them – probably in their attempts to include Iran, Indonesia, and now Nigeria – cannot be trusted. The problem before the Europeans, therefore, is to stockpile the oil so that if total secession of oil supply were to take place, stockpiling would determine who would stay longer in the race. If Adelman has remembered that Iran – the most vulnerable state among the oil-producing countries from the standpoint of foreign exchange – had, at one time, resisted an economic siege for over three years, he would have rethought his stockpiling doctrine.

One can equally argue that the Europeans' security problem is in fact functional to their oil supply being dominated by huge Anglo-Saxon companies. However, one would hasten to assert that 'xenophobia' is not the issue here. The Europeans are worried over their security because they know that the companies which control their oil supply belong to a country which is in open support of a state which is at war and occupying lands in the countries of their oil supply sources. The Europeans certainly know that the Arabs have no special liking to hurt them. They know, for example, that oil became a political problem and that the TAP Line remained broken in the Syrian desert for over nine months because it is controlled by America, which supports Israel. This, at least, is what one would ascertain from the testimony on the Near East conflict by Charles C. Bonin, President of the American Arab Association for Commerce and Industry, before the Subcommittee on the Near East of the Committee on Foreign Affairs of the House of Representatives. Bonin testified:

> Specifically for us in the United States, we believe that our policy in the Middle East should be based upon our national interest in all its dimensions – strategic, economic, and political. In view of the increased importance of the Soviet

role in the Middle East, we consider it of the utmost importance to arrest the trend toward polarization of the region into Western and Soviet spheres of influence. For if this trend should continue, it will not be too long before our allies in Western Europe and the Far East will be confronted with the necessity of accommodating themselves to the resulting situation in order to safeguard their own interests. Inevitably, those countries which depend for 55 to 90 percent of their energy needs upon the resources of the Arab world will turn to the Arab countries and their allies to guarantee the security of their energy requirements. Such a development will accelerate the waning of American influence in that region and beyond, with incalculable consequences for our own security.[49]

This is certainly what had been in the minds of the officials of Italy's state oil firm, ENI, when they refused to associate themselves with the New York declaration discussed earlier. An official from the Italian company said, 'ENI feels that it is committed to protect interests different from those of the international oil companies.' The official went on to declare that 'there exist considerable differences between them in the manner of understanding competition in the world and Italian energy markets, and on the type of cooperative relations to establish with oil-producing countries.'[50]

It is this context that one should view the growth of European oil companies reaching for direct deals with the national oil companies of the oil-producing countries. Almost every European country now considers it in its national interest to have its oil company in possession of its energy at the source.[51] This has the advantage of identifying an oil-consuming country's interest with the interest of an oil-producing one, thus making both countries less susceptible to capricious behavior. It would also have the advantage of disengaging the oil business from the repercussions of a totally different set of occurrences. Thus a Danish woman would no longer be tortured by spending a cold night in an unheated room because a Palestinian woman had been uprooted from her home town, or because a Mosque was set on fire.

The government-to-government approach had been pioneered by ENI, the Italian company. In 1957, ENI and the Iranian Oil Company formed a joint venture, SIRIP, which greatly upset the then prevailing producing countries to companies relations. The agreement called for a tax rate of 50 percent on SIRIP which, in effect, gave the Iranian government 75 percent of the net income of the company.[52] The process of direct deals had been accelerated by the rise of a European consortium from the French government company, ELF-ERAP; the Italian company, ENI; the Spanish government company, Hispanoil; the Belgian government company, Petrofira; and the Austrian oil company, OMV. The purpose of this consortium is to develop more secure oil for their own countries.[53] Government-to-government deals have also been gaining momentum by the avowed intention of the officials in charge of the energy question in the European Community to have 'direct cooperation with the oil exporting countries' through bilateral deals covering trade, know-how, and extension of credits.[54]

This then brings us to the long-term objective in OPEC's list – participation. As noted earlier, the idea of participation had been recently introduced by the Italian oil company, ENI. The pattern of participation followed is to have a foreign company enter into a joint enterprise with a national oil company of a producing country. The territories covered by the new participation arrangements included areas relinquished by old companies or offshore areas. As of now National Iranian Oil Company has concluded at least eight participant agreements with Italian companies, American independents, Spanish, French, and German oil companies. Similarly, Saudi Arabian General Petroleum and Mineral Organization (Petromin) had in 1965 concluded with AUXIRAP, a French state-owned company, a participant agreement for oil exploration in an offshore area in the Red Sea. Also, in 1967 Petromin concluded another participant agreement with two American independent oil companies. A third agreement was concluded with the Italian state-owned company, AGIP, for oil exploration in the southeast territory of the country. In 1967, Kuwait National Oil Company concluded a participant agreement with the Spanish state-owned oil company, Hispanoil.

None of these agreements touched upon the major concession-holding companies. OPEC's long-standing goal of state participation in the ownership of existing concessions continued to be addressed in almost every OPEC gathering. In its twenty-fourth conference, held in Vienna in July 1977, OPEC reiterated its faith in the principle of participation as the wave of the future. Then at a special conference held in Beirut on September 22, OPEC resolved that all member countries concerned shall establish negotiations with the oil companies either individually or in groups with a view to achieving participation. If such negotiations fail, said the resolution, the next conference will determine the step to be taken 'with a view to enforce and achieve the objective of effective participation.'[55] The Organization then established a special ministerial committee to undertake detailed study for the implementation of effective participation in the ownership, management, and staffing of the companies.[56]

Some OPEC members, Algeria for one, had achieved control over the oil industry through nationalization. While nationalization would give a country the desired control, it, nevertheless, is usually accompanied by the rupture of country–company relations, and the ensuing conflicts could lead to a loss of markets. Participation, on the other hand, is a system designed to give a country effective control over the oil industry through harmonious relations with its concessionaire company. It is founded on the conclusion of agreements to effect the transfer of ownership and to prevent harmful competition. The course OPEC chooses to follow to bring about effective control over the oil industry could, therefore, smoothly change the future of the major oil companies in a profound and permanent way. Participation, if carried further, could turn the major oil companies into a group of employees of the oil-producing countries.

If Libya triggered the price negotiations and Algeria pioneered the first successful nationalization, it was Saudi Arabia who championed the cause of participation. Saudi Arabia had always been behind the principle of participation in existing concessions. Accordingly, OPEC chose in January 1972, the Saudi Arabian Minister for Petroleum and Mineral Resources to commence negotiations with the companies on behalf of the Gulf oil-producing states on the issue of participation. It was then natural for the Saudi Arabian minister to start the negotiations with the major oil group in his own country, i.e., the Arabian American Oil Company (Aramco). Negotiations between the two parties commenced, only to reach an impasse by the end of February, 1972. It was at this juncture that King Faisal of Saudi Arabia, who was taking a direct interest in conducting the negotiations, sent a personal message to the oil companies demanding satisfactory agreement on the participation issue. In his message the King declared that 'The implementation of effective participation is imperative, and we expect the companies to cooperate with us with a view to reaching a satisfactory agreement.' The King went on to warn that 'they should not oblige us to take measures in order to put into effect the implementation of participation.'[57]

Thus, on March 10, 1972, while OPEC member countries were attending their special ministerial session in Beirut, Aramco handed the Saudi Arabian minister its acceptance of the principle of 20 percent government participation. Aramco's action was then followed by all the companies in OPEC member countries. Thus, OPEC's primary goal, first established in Resolution XVI.90 on June 24-25, 1968, finally materialized.[58]

Admittedly, there exists a long road to travel to reach a settlement over the issue of compensation. OPEC members are now talking of paying cash for the net book value of above-ground assets. On the other hand, companies are holding out for complete compensation for the loss of future profits from the share of oil production they surrendered. Eventually, an agreement will be reached. Talks have already begun between Aramco and officials in Saudi Arabia with the intention of reaching a settlement before OPEC's next conference in June, 1972. Such settlement is expected to serve as a blueprint for OPEC's other members.

The question now is whether this is a phased-out participation. At least, every official of the oil-producing countries is on record as declaring that the 20 percent participation is just the beginning. OPEC member countries had on a number of occasions stated that the intention is to have a participating share of 51 percent and eventually 100 percent as existing concessions expire.

These processes indicate that the oil-producing countries, already in effective control of their oil resources, will very soon achieve a majority ownership of those resources.

One looks at these developments and observes that while rivalry among the superpowers is dominating the scene, there forges ahead a cluster of oil-producing and oil-consuming power centers bent on grasping the control for themselves.

Endnotes

[1] Organization of Petroleum Exporting Countries (OPEC), *Background Information*, Geneva, 1964, 5.

[2] Aliric Parra, 'Production Pressures and Parameters, Inter-Relations of Countries and Companies,' *The Princeton University Conference and Twentieth Annual Near East Conference on Middle East Focus: The Persian Gulf, October 24-25, 1968*, Cuyler Young (ed.) (Princeton: The Princeton University Conference, 1968), 76.

[3] OPEC, *Background Information*, op. cit., 11.

[4] OPEC, Speech Delivered by Mr. Faud Rouhani, Secretary General, at II Consultative Meeting, Geneva, July, 1963, 4.

[5] OPEC, *Background Information*, op. cit., 11–14.

[6] Frank Gardner, 'International Oil Beset on Every Side,' *Oil and Gas Journal*, LXIX, No. 3 (January 18, 1971), 35.

[7] OPEC, *Background Information*, op. cit., 15.

[8] Ibid., 12.

[9] Ibid., 13.

[10] OPEC, *The Oil Industry's Organization in the Middle East and Some of Its Fiscal Consequences*, by F.R. Parra, Geneva, November, 1963, 9; 'When Do the Concessions End?' *Petroleum Press Service*, XXXVIII, No. 12 (December, 1971), 449.

[11] OPEC, *Exporting Countries and International Oil*, London, May 1964, 9.

[12] OPEC, *The Oil Industry's Organization in the Middle East and Some of Its Fiscal Consequences*, op. cit., 11.

[13] OPEC, *OPEC and the Principle of Negotiation*, presented to the 5th Arab Petroleum Congress, Cairo, March 16-23, 1965, 18.

[14] See Bertram B. Johansson, 'Bad Feeling Over Oil Hikes, Libyan Housecleaninq Gets Few Plaudits From Abroad,' *Christian Science Monitor* (May 26, 1971), 4.

[15] Editorial, 'A Successful Bargaining Scheme,' *World Petroleum*, XLI, No. 10 (October, 1970), 21; Gene T. Kinney, 'Nixon Moves to Aid Foreign Oil Talks,' *Oil and Gas Journal*, LXIX, No. 4 (January 25, 1971), 84-85; 'Consumers Pay the Price,' *Petroleum Press Service*, XXXVIII, No. 4 (April 1971), 124.

[16] OPEC, *Background Information*, op. cit., 16.

[17] OPEC, *OPEC and the Principle of Negotiation*, op. cit., 8-9.

[18] OPEC, *OPEC and the Principle of Negotiation*, op. cit., 17.

[19] OPEC, *Resolutions of the Ninth Conference*, Tripoli, July 7-13, 1965.

[20] 'OPEC Gets to Essentials,' *The Economist*, CCXVI (August 28, 1965), 801.

[21] OPEC, *1967 Review and Record*, Vienna, 1967, 4.

[22] Walter J. Levy, 'Oil Power,' *Foreign Affairs*, XLIX, No. 4 (July, 1971), 652.

[23] OPEC, *Annual Review and Record*, 1968, 17-22.

[24] Levy, op. cit., 652.

[25] 'OPEC Seeks A Plan,' *Petroleum Press Service*, XXXVII, No. 9 (September, 1970), 318-320; 'OPEC Strongly Backs All Arab Claims,' *Oil and Gas Journal*, LXVIII, No. 32 (August 10, 1970), 78-79.

[26] Gurney Breckenfeld, 'How the Arab Changed the Oil Business,' *Fortune*, LXXXIV, No. 2 (August, 1971), 116; 'Oil Production Still Restricted,' *Middle East Economic Digest*, XIV, No. 25 (June 19, 1970), 736; 'Libya Pressuring Concessionaires to Boost Drilling,' *Oil and Gas Journal*, LXVIII, No. 23 (June 8, 1970), 63.

[27] *Middle East Economic Digest*, XIV, No. 23 (June 5, 1970), 680; *Middle East Economic Digest*, XIV, No. 29 (July 17, 1970), 849-850; *Middle East Economic Digest*, XIV, No. 30 (July 24, 1970), 879.

[28] *Middle East Economic Digest*, XIV, No. 33 (August 14, 1970), 963; *Middle East Economic Digest*, XIV, No. 35 (August 28, 1970), 1020.

[29] *Middle East Economic Digest*, XIV, No. 36 (September 4, 1970), 1047; Editorial, 'A Successful Bargaining Scheme,' *World Petroleum*, op. cit., 21.

[30] *Middle East Economic Digest*, XIV, No. 37 (September 11, 1970), 1076; *Middle East Economic Digest*, XIV, No. 40 (October 2, 1970), 1157; *Middle East Economic Digest*, XIV, No. 43 (October 23, 1970), 1243; Editorial, 'A Successful Bargaining Scheme,' *World Petroleum*, op. cit., 21.

[31] *Middle East Economic Digest*, XIV, No. 40 (October 2, 1970), 1155; *Middle East Economic Digest*, XIV, No. 41 (October 9, 1970), 1184.

[32] *Middle East Economic Digest*, XIV, No. 44 (October 30, 1970), 1265; John K. Cooley, 'French Take Spill on Algerian Oil,' *Christian Science Monitor* (April 16, 1971), 1.

[33] John K. Cooley, 'Algerian-Libyan Common Oil Front Faces Solidarity Test Soon?' *Christian Science Monitor* (April 30, 1971), 15.

[34] 'Defense, State Okay Imports of Algerian LNG,' *Oil and Gas Journal*, LXIX, No. 31 (August 2, 1971), 29.

[35] 'OPEC Working for Even Bigger Share,' *Oil and Gas Journal*, LXIX, No. 1 (January 4, 1971), 48.

[36] 'Iran Threatens Oil Stoppage,' *Christian Science Monitor* (January 26, 1971), 2.

[37] Gene T. Kinney, 'Nixon Moves to Aid Foreign Oil Talks,' *Oil and Gas Journal*, LXIX, No. 4 (January 25, 1971), 84-85.

[38] 'Oil Firms Team Up on Mideast,' *Christian Science Monitor* (January 20, 1971), 1-2; Frank J. Gardner, 'OPEC Faced With Collective Bargaining,' *Oil and Gas Journal*, LXIX, No. 4 (January 25, 1971), 82-83.

[39] Text in *The Middle East and North Africa, 1971-72* (18th edn.; Europa Publications Limited, 1971), 37.

[40] Ibid.

[41] Frank J. Gardner, 'Libyan Oil Agreement Makes Big Waves,' *Oil and Gas Journal*, XLIX. No. 15 (April 12, 1971), 32.

[42] Frank J. Gardner, 'Algerians Post $3.60/BBL Crude Price,' *Oil and Gas Journal*, LXIX, No. 16 (April 19, 1971), 90.

[43] 'Persian Gulf Crude Prices Jump Again,' *Oil and Gas Journal*, LXIX, No. 24 (June 14, 1971), 44; John K. Cooley, 'New Iraq Accord Nearly Doubles Oil Income,' *Christian Science Monitor* (June 24, 1971), 3.

[44] Quoted in 'Shah: Eliminate Oil Production System,' *Middle East Economic Digest*, XIV, No. 25 (June 19, 1970), 731.

[45] M.A. Adelman, 'Security of Eastern Hemisphere Fuel Supply,' Working Paper No. 6, Department of Economics,

(Cambridge: Massachusetts Institute of Technology, December, 1967), 3.

⁴⁶ Ibid., 5.

⁴⁷ 'Cheapness With Security,' *Petroleum Press Service*, XXXV, No. 1 (January, 1968), 2-4.

⁴⁸ See Frank J. Gardner, 'World Oil Industry Licks Its Wounds, Plans Ahead,' *Oil and Gas Journal*, LXIX, No. 25 (June 21, 1971), 73.

⁴⁹ US Congress, House, Subcommittee on the Near East of the Committee on Foreign Affairs, *The Near East Conflict, Hearings*, 91st Cong., 2nd Sess., July 21, 22, 23, 28, 29, and 30, 1970 (Washington DC: Government Printing Office, 1970), 207.

⁵⁰ Quoted in Frank J. Gardner, 'OPEC Faced with Collective Bargaining,' *Oil and Gas Journal*, op. cit., 82-83; John K. Cooley, '10-Power Oil Talks Set Precedent,' *Christian Science Monitor* (January 21, 1971), 9.

⁵¹ 'Europe's National Oil Companies,' *Petroleum Press Service*, XXXVI, No. 11 (November, 1969), 414.

⁵² Christopher Tugendhat, *Oil: The Biggest Business* (New York: G.P. Putnam's Sons, 1968).

⁵³ 'Europe's National Oil Companies,' *Petroleum Press Service*, op. cit., 414.

⁵⁴ 'Towards Bilateral Arrangements,' *Petroleum Press Service*, XXXVIII, No. 5 (June, 1971), 189.

⁵⁵ 'Oil Unit Demands A Say,' *Christian Science Monitor* (October 12, 1971), 5.

[56] 'OPEC Stepping Up State-Participation Drive,' *Oil and Gas Journal*, LXIX, No. 31 (August 2, 1971), 38.

[57] Quoted in Frank J. Gardner, 'A King Speaks,' *Oil and Gas Journal*, LXX, No. 9 (February 26, 1972), 27; 'OPEC to Ponder New Participation Jolt,' *Oil and Gas Journal*, LXX, No. 11 (March 13, 1972), 47.

[58] 'Aramco to Grant Direct 20% Stake to Saudi Arabia,' *Wall Street Journal* (March 13, 1972), 4; Frank J. Gardner, 'Persian Gulf Oils Bow to Participation,' *Oil and Gas Journal*, LXX, No. 13 (March 27, 1972), 47-49.

1

Conclusion

The cheapest and most abundant oil supplies are concentrated in the Persian Gulf area. Corollary to this concentration is European and Japanese heavy dependence on these oil resources. Also, it is estimated that by the end of the 1970s the United States will be importing oil from the Gulf region and North Africa on the order of one-half of her petroleum needs, and that the Soviet bloc will be experiencing a 'fuel deficit' of more than 100 million tons per year which will be filled by imports from the developing countries.[1]

The Gulf States, therefore, continue to increase in importance and be the scene of a renewed East–West rivalry; for the control of these states and their resources brings with it an important leverage in the bidding for world leadership. Inevitably, the control of these oil resources will force upon the oil-consuming countries in Europe and Japan an accommodation to the directives of the controller.

Today, the United States is in control of the larger portion of the oil in the Gulf. In addition to its being an impressive source of financing the United States' balance of payments deficits, oil in the Gulf, controlled by the Americans, is a factor in the exercise of world leadership.

The American position, however, has been challenged by the rising power of the Soviet Union. After the British military departure from the Gulf area, the United States appeared reluctant to beef up its token naval detachment on the island of Bahrain. She preferred to watch a Soviet drive towards the Gulf from a communication base on the island of Diego Garcia, out in the wide reaches of the Indian Ocean, rather than be involved in the Gulf politics, and, hence, be charged with pursuing an imperialist policy.

This military stand did save the United States from a renewal of the imperialist charge, but it could not make up for the deteriorating position the United States now suffers in the Arab world.

In their bidding for the Jewish vote (the strength of which is questionable since both Presidents Eisenhower and Nixon assumed the presidency without its support), American leaders fell into the dilemma of pursuing two opposing objectives: nurturing the United States' interest in the Arab world and supporting the state of Israel. Needless to say, the United States opted for the latter objective.

This stand has facilitated the task of the Soviet Union in her drive towards the Gulf. This Soviet move, however, should be viewed as its being a link in a long chain of Soviet global strategy aiming to control oil for its significance in affecting a change in the political orientation of Europe and Japan.

Taking advantage of America's foreign policy which resulted in strained Arab-American relations, the Soviet Union leaped over Turkey, Iran and Pakistan – the northern tier – and concentrated its efforts on Egypt first, and then Syria, Yemen, South Yemen, and Iraq. The countries of the northern tier, witnessing the Soviets outflank them, saw no advantage in holding onto their alliance with the West – in anticipation of a Soviet drive from the north – while the Soviets were already in their backyards. It became ridiculous for Iran and Turkey to continue their refusal of cooperation with the Soviet Union. Hence the extensive Iran–Russian cooperations and the normalization of Soviet-Turkish relations, the latest sign of which was the visit made by President Nikolai Podgorney to the Turkish capital on April 11-18, 1972.[2]

With their task nearly completed with success in the northern tier, the Russians found the Persian Gulf amenable to providing them with their age-long need for warm water, and the instrument necessary for influencing the political behavior of the NATO members and Japan. In one leap, the Russians had most of the Arab world, the northern tier, and now Europe and Japan either in an open quarrel with the United States or breaking away from American influence.

Specifically, four steps had been taken by the Russians in pursuing their objectives in the Gulf area. First, they embarked on a policy of presenting new options to the oil-producing countries through the conclusion of oil and gas agreements, thus enabling them to break away from western influence.

Second, the Russians resorted to making overt commitments to the countries in the areas of their activities. Thus, after the signing of treaties with Egypt and India, the Soviet Union signed on April 10, 1972, a 15-year treaty with Iraq, a treaty which marked the inauguration of the North Rumaila oilfields, developed with Soviet aid. Premier Alexei N. Kosygin attended the ceremony and in his speech he did not forget to promise the Arab people to 'free their wealth' from 'Western monopolies.'[3]

Third, in support of this policy, the Russians opted for a policy of expanding their Indian Ocean fleet. Soviet naval vessels are becoming the instrument for the extension of Soviet political influence through their periodic calls on ports in the Persian Gulf.

Finally, in their attempts to weaken Western influence, the Soviets are rivaling the Chinese in support of guerrilla warfare by the Popular Front for the Liberation of the Occupied Arab Gulf. The front, based in the province of Dhufar of the Sultanate of Oman, provided the ammunition and training from across the border of the People's Republic of Yemen. The intention of the front is the liquidation of Western presence in the Gulf area.

Ironically, these impressive Soviet achievements are often belittled in some American quarters as being important but not necessarily detrimental to American interests. Those Americans should be reminded, however, of the Soviets' precedents in the Mediterranean, on the Suez Canal, at Aden, and in Iraq. They should be reminded that it was a Soviet commitment to India which reduced Pakistan from a country of 120 million people to a middle-sized state.

This, then, is the rivalry which is now dominating the scene in the Persian Gulf. But while this rivalry is taking place between the superpowers, there emerges a power center from the oil-producing countries which makes use of this superpower struggle. The oil-producing countries gathered together in the formation of the Organization of Petroleum Exporting Countries (OPEC). Recognition of a common interest among these countries has made it possible for them to have a say in determining the price at which oil gets sold, in effectuating a change in the tax rate, and in acquiring minority participation in the ownership of the industry. It will not be very long before these countries receive a majority ownership of their resources.

Ownership, however, is different from controllership. Who gets what, especially in an emergency, will be decided by political orientation and political expediency. American foreign policy has made it possible for the Soviets to dig into the Arab world. The trend indicates that the Soviets are drilling deep in the Persian Gulf. Very soon they will help in the development of certain political orientation in the oil-producing countries of the area which they will then use as vehicles to promote Soviet political expediency. At the same time Europe and Japan will continue to receive oil, but through a Russian pipe. And no other states will be powerful enough or aware enough to attempt reverse the trend.

Endnotes

[1] US, Congress, House, Subcommittee on the Near East of the Committee on Foreign Affairs, *The Middle East, 1971: The Need to Strengthen the Peace Hearings*, 91st Cong., 1st Sess., July 13, 14, 15, 27; August 3; September 30; October 5 and 28, 1971 (Washington DC: Government Printing Office, 1971), 20; 'Soviet Interest in Middle East Oil,' *Mizan*, XIV, No. 1 (August, 1971), 30-33.

[2] John K. Cooley, 'Soviets Push Interests in Asia,' *Christian Science Monitor* (April 12, 1972), 1 and 11.

[3] Quoted in ibid., 11.

Bibliography

Official Documents

Great Britain. Foreign Office. *Documents on British Foreign Policy 1919-1939*. E.L. Woodward and Rohan Butler, eds. Vol. IV, 1st Series. London, 1952.

Great Britain. Parliament. *Parliamentary Debates, House of Commons*. Vol. CCCLXXXIV, September 30, 1941.

_____. *Parliamentary Debates, House of Commons*. CCCCLXXXIX, Fifth Series, June 21, 1951.

_____. League of Nations. *Official Journal*. 13th Year, Vol. XIII². 1932.

_____. *Official Journal*. Vol. XIV¹. February, 1933.

_____. *Treaty Series*. Vol. IX. 1922.

The Organization of Petroleum Exporting Countries (OPEC). *Annual Review and Record*. Vienna, 1968.

_____. *Background Information*. Geneva, 1964.

_____. *Demand Patterns and Crude Gravities*, by F.R. Parra. Geneva, 1963.

_____. *The Development of Petroleum Resources Under the Concession System in Non-Industrialized Countries*, by F.R. Parra. Geneva, 1964.

_____. *Elasticity of Demand for Crude Oil; Its Implications for Exporting Countries*, by I.K. Kabbani. Geneva, May, 1964.

_____. *Exporting Countries and International Oil*. London, May, 1964.

_____. *Explanatory Memoranda of the Memoranda of the OPEC Resolutions*. Geneva, April-June, 1962.

_____. *From Concession to Contracts*. Cairo, March 1, 1965.

_____. *The Oil Industry's Organization in the Middle East and Some of its Fiscal Consequences*, by F.R. Parra. Geneva, November, 1963.

_____. *OPEC Bulletin*. September-October, 1969.

_____. *OPEC and the Principle of Negotiation*. Cairo, 1965.

_____. *Pricing Problems: Further Considerations*. Geneva, September, 1963.

_____. *Radical Changes in the International Oil Industry During the Past Decade*. Beirut, November, 1963.

_____. *Resolutions of the Ninth Conference*. Tripoli, July 7-13, 1965.

_____. *1967 Review and Record*. Vienna, 1967.

_____. *Speech Delivered by Mr. Fuad Rouhani, Secretary General at II Consultative Meeting*. Geneva, July, 1963.

_____. *Taxation Economics in Crude Production*. Cairo, 1965.

United Nations. Security Council. *Official Records*. 1st Year, 1st Series, No. 2, March-June, 1946.

_____. *Official Records*. 1st Year, 1st Series, Supplement No. 1, January-February, 1946.

_____. *Official Records*. Sixth Year, 563rd Meeting, October 17, 1951.

US Congress. House. Subcommittee on Mines and Mining of the
Committee on Interior and Insular Affairs. *Oil Import
Controls, Hearings*, Serial No. 91-17, 91st Cong., 2nd Sess.,
March 9, 10, 16 and 17, April 6, 7, 23 and 24, 1970.
Washington DC, Government Printing Office, 1970.

_____. Subcommittee on the Near East of the Committee on Foreign
Affairs. *The Middle East, 1971: The Need to Strengthen the
Peace Hearings.* 91st Cong., 1st Sess., July 13, 14, 15, 27;
August 3, September 30, October 5 and 28, 1971.
Washington DC, Government Printing Office, 1971.

_____. Subcommittee on the Near East of the Committee on Foreign
Affairs. *The Near East Conflict, Hearings.* 91st Cong., 2nd
Sess., July 21, 22, 23, 28, 29 and 30, 1970. Washington DC,
Government Printing Office, 1970.

US Congress. Senate. *Oil Concession in Foreign Countries.* Doc. No.
97. 68th Cong., 1st Sess. Washington DC, Government
Printing Office, 1924.

_____. Special Committee Investigating the National Defense
Program. *Petroleum Arrangements with Saudi Arabia,
Hearings.* Part 41. 80th Cong., 1st Sess., March 28, 29; May 8;
October 29, 30, 31; November 1, 3, 4, 1947 and January 24,
29, 30, 1948. Washington DC, Government Printing Office,
1948.

_____. Special Committee Investigating Petroleum Resources.
American Petroleum Interests in Foreign Countries, Hearings.
79th Cong., 1st Sess., June 27 and 28, 1945. Washington DC,
Government Printing Office, 1946.

_____. Subcommittee on Antitrust and Monopoly of the Committee
on Judiciary. *The Petroleum Industry, The Cabinet Task Force
on Oil Import Control: Majority and Minority
Recommendations, Hearings.* Part 4. 91st Cong., 2nd Sess.,

March 3 and 26, 1970. Washington DC, Government Printing Office, 1970.

US Congress. Senate and House. Select Committee on Small Business of the Senate and the House of Representatives. *A Report: The Third Petroleum Congress*. 82nd Cong., 2nd Sess. Washington DC, Government Printing Office, 1952.

US Department of Interior. Bureau of Mines. 'The Mineral Industry of the USSR,' *Minerals Yearbook, Area Reports: International*, IV. Washington DC, Government Printing Office, 1971.

US Department of State. *Bulletin*. Vol. 24, No. 616, April 3, 1951.

_____. *Bulletin*. Vol. 24, No. 616, April 23, 1951.

_____. *Bulletin*. Vol. 25, No. 628, July 9, 1951.

_____. *Bulletin*. Vol. 29, No. 773, July 13, 1953.

_____. *1972-1994 Voting Practices in the United Nations Bureau of International Organization Affairs*. Report to Congress. Submitted Pursuant to Public Law 101-167, March 31, 1995.

_____. *Lessons to be Learned from the 66 U.N. Resolutions Israel Ignores*. Washington Report on Middle East Affairs, Washington DC, 1993.

_____. *Papers Relating to the Foreign Relations of the United States, 1920*. Vol. 3. Washington DC, Government Printing Office, 1936.

_____. *Papers Relating to the Foreign Relations of the United States, 1921*. Vol. 2. Washington DC, Government Printing Office, 1936.

_____. *Papers Relating to the Foreign Relations of the United States,*

1923. Vol. 2. Washington DC, Government Printing Office, 1938.

_____. *Papers Relating to the Foreign Relations of the United States, 1925*. Vol. 2. Washington DC, Government Printing Office, 1940.

_____. *Papers Relating to the Foreign Relations of the United States, 1927*. Vol. 2. Washington DC, Government Printing Office, 1942.

_____. *Papers Relating to the Foreign Relations of the United States, 1929*. Vol. 3. Washington DC, Government Printing Office, 1944.

_____. *Papers Relating to the Foreign Relations of the United States, 1932*. Vol. 2. Washington DC, Government Printing Office, 1947.

US Federal Trade Commission. *The International Petroleum Cartel, Staff Report Submitted to the Subcommittee on Monopoly of the Select Committee on Small Business.* 82nd Cong., 2nd Sess., No. 6. Washington DC, Government Printing Office, 1952.

Books

Amin, Abdul Amir. *British Interests in the Persian Gulf.* Leiden, Netherlands: E.J. Brill, 1967.

Anderson, J.R.L. *East of Suez: A Study of Britain's Greatest Trading Enterprise.* London: Hodder and Staughton, 1969.

Bartholomew, J.G. *The Times Survey Atlas of the World.* London: The Times, 1922.

Bullard, Sir Reader, ed. *The Middle East: A Political and Economic*

Survey. London, New York, and Toronto: Oxford University Press, 1958.

Busch, Briton Cooper. *Britain and the Persian Gulf, 1894-1914*. Berkeley and Los Angeles: University of California Press, 1967.

Campbell, Robert W. *The Economics of Soviet Oil and Gas*. Baltimore: The Johns Hopkins Press, 1968.

Caroe, Olaf. *Wells of Power, the Oilfields of South-Western Asia, A Regional and Global Study*. London: Macmillan and Co., Ltd., 1951.

The Center for Strategic and International Studies. *The Gulf: Implications of British Withdrawal*. Special Report, Series No. 8. Washington DC: Georgetown University, 1969.

Christopher, W., H. Saunders, G. Sick and P. Kriesberg. *American Hostages in Iran: The Conduct of a Crisis*. New Haven: Yale University Press, 1985.

Coke, Richard. *The Arab's Place in the Sun*. London: Thornton Butterworth Ltd., 1929.

DeNovo, John. *American Interests and Policies in the Middle East: 1900-1939*. Minneapolis: The University of Minnesota Press, 1963.

Ellis, Harry B. *Challenge in the Middle East: Communist Influence and American Policy*. New York: The Ronald Press Company, 1960.

Engler, Robert. *The Politics of Oil: A Study of Private Power and Democratic Direction*. New York: The Macmillan Company, 1961.

Finnie, David H. *The Middle East Oil Industry in Its Local Environment*. Cambridge: Harvard University Press, 1958.

Fitzsimmons, M.A. *Empire by Treaty: Britain and the Middle East in the Twentieth Century*. Notre Dame, Indiana: University of Notre Dame Press, 1964.

Gibb, G.S. and E.H. Knowlton. *The Resurgent Years, History of Standard Oil Company (New Jersey) 1911-1927*. New York: Harper & Bros., 1956.

Hamilton, Charles W. *Americans and Oil in the Middle East*. Houston, Texas: Gulf Publishing Company, 1962.

Hardinge, Sir Arthur H. *A Diplomatist in the East*. London: J. Cape Limited, 1928.

Hay, Sir Rupert. *The Persian Gulf States*. Washington DC: The Middle East Institute, 1959.

Hoskins, Halford L. *The Middle East: Problem Area in World Politics*. New York: The Macmillan Company, 1954.

Howarth, David. *The Desert King: Ibn Saud and His Arabia*. New York, Toronto, London: McGraw-Hill Book Company, 1964.

Hull, Cordell. *The Memoirs of Cordell Hull*, Vol. II. New York: The Macmillan Company, 1948.

Hunter, Robert E. *The Soviet Dilemma in the Middle East. Part II: Oil and the Persian Gulf*. Adelphi Papers, No. 60. London: Institute for Strategic Studies, 1969.

Hurewitz, J.C. *Middle East Dilemmas: The Background of United States Policy*. New York: Harper and Brothers, 1953.

_____, ed. *Soviet-American Rivalry in the Middle East*. New York,

Washington, London: Frederick. A Praeger, Publishers, 1969.

Kelly, J.B. *Eastern Arabian Frontiers*. London: Faber and Faber, 1964.

Kirk, George Edwards. *The Middle East in the War, Survey of International Affairs 1939-1946*. London: Oxford University Press, 1952.

Kumar, Ravinder. *India and the Persian Gulf Region 1858-1907: A Study in British Imperial Policy*. New York: Asia Publishing House, 1956.

Laqueur, Walter Z. *Communism and Nationalism in the Middle East*. New York: Frederick A. Praeger, Publishers, 1956.

_____. *The Struggle for the Middle East: The Soviet Union in the Mediterranean 1958-1968*. New York: The Macmillan Company, 1969.

Leeman, Wayne A. *The Price of Middle East Oil: An Essay in Political Economy*. Ithaca, New York: Cornell University Press, 1962.

Lenczowski, George. *The Middle East in World Affairs*. 3rd edn. Ithaca, New York: Cornell University Press, 1962.

_____. *Oil and State in the Middle East*. Ithaca, New York: Cornell University Press, 1960.

_____, ed. *United States Interests in the Middle East*. Washington DC: American Enterprise Institute for Public Policy Research, 1968.

Lewis, Bernard. *The Middle East and the West*. Bloomington: Indiana University Press, 1964.

Longrigg, Stephen Hemsley. *Oil in the Middle East: Its Discovery and*

Development. 2nd edn. London: Oxford University Press, 1961.

The Middle East and North Africa, 1969-70. 16th edn. London: Europa Publications Limited, 1969.

The Middle East and North Africa, 1970-71. 17th edn. London: Europa Publications Limited, 1970.
The Middle East and North Africa, 1971-72. 18th edn. London: Europa Publications Limited, 1971.

Mikdashi, Zuhayr. *A Financial Analysis of Middle Eastern Oil Concessions: 1901-1965*. New York, Washington and London: Frederick A. Praeger, Publishers, 1965.

Mikesell, Raymond F. and Hollis B. Chenery. *Arabian Oil: America's Stake in the Middle East*. Chapel Hill: The University of North Carolina Press, 1949.

Miles, S.B. *The Countries and Tribes of the Persian Gulf*. London: Frank Cass and Co. Ltd., 1966.

Millspaugh, Arthur Chester. *Americans in Persia*. Washington DC: The Brookings Institution, 1946.

Motter, T.H. Vail. *U.S. Army in World War II, The Middle East Theater: The Persian Corridor and Aid to Russia*. Office of the Chief Military History, Department of the Army. Washington DC: Government Printing Office, 1952.

National Bank of Egypt. *The International Oil Industry in the Middle East*. By Edith Penrose. Cairo, 1968.

Owen, Roderic. *The Golden Bubble: Arabian Gulf Documentary*. London and Glasgow: Collins Clear-Type Press, 1957.

Peretz, Don. *The Middle East Today*. New York, Chicago, San

Francisco, Toronto, London: Holt, Rinehart and Winston, Inc., 1963.

Sanger, Richard H. *The Arabian Peninsula*. Ithaca, New York: Cornell University Press, 1954.

Sharabi, H.B. *Governments and Politics of the Middle East in the Twentieth Century*. Princeton, Toronto, London, and New York: D. Van Nostrand Company, Inc., 1962.

Shwadran, Benjamin. *The Middle East, Oil and the Great Power*. New York, Frederick A. Praeger, Publishers, 1955.

Standard Oil Company (New Jersey). *A Background Memorandum on Company Policy and Actions*. New York: Standard Oil Company, 1947.

Thompson, Jack H. and Robert O. Reischauer, eds. *Modernization of the Arab World*. Princeton, Toronto, New York, and London: D. Van Nostrand Company, Inc. 1966.

Tugendhat, Christopher. *Oil: The Biggest Business*. New York: G.P. Putnam's Sons, 1968.

Walpole, Norman C. *Area Handbook for Saudi Arabia*. Washington DC: Government Printing Office, 1966.

Wilson, Sir Arnold T. *The Persian Gulf: An Historical Sketch from the Earliest Times to the Beginning of the Twentieth Century*. London: George Allen and Unwin, Ltd., 1928.

Yergin, Daniel. *The Prize*. New York: Touchstone, 1991.

Young T. Cuyler, ed. *The Princeton University Conference and Twentieth Annual Near East Conference on Middle East Focus: The Persian Gulf, October 24-25, 1968*. Princeton: The Princeton University Conference, 1968.

Articles

Adelman, M.A. 'Security of Eastern Hemisphere Fuel Supply,'
Working Paper No. 6, Department of Economics,
Massachusetts Institute of Technology, Cambridge,
Massachusetts (December, 1967).

'Aramco to Grant Direct 20% Stake to Saudi Arabia,' *Wall Street Journal* (March 13, 1972).

Ashworth, George W. 'Russians in the Indian Ocean: Assessment
from Washington,' *Christian Science Monitor* (November 17,
1970).

Bachman, W.A. 'Is the US Vastly Underestimating Its Oil Needs?'
Oil and Gas Journal, LXIX, No. 8 (February 22, 1971).

'Basic Analysis, Oil,' *Standard and Poor's Industry Surveys*, Section 2
(December 11, 1969).

'Basic Analysis, Oil,' *Standard and Poor's Industry Surveys*, Section 4
(April 30, 1970).

Bayne, Edward Ashley. 'Crisis of Confidence in Iran,' *Foreign Affairs*, XXIX, No. 4 (July, 1951).

'Big Rise in Europe's Imports,' *Petroleum Press Service*, XXXVIII, No.
5 (May, 1971).

Boyce, Richard H. 'Mideast Peace Settlement Eyed,' *Rocky Mountain News* (May 21, 1971).

Breckenfeld, Gurney. 'How the Arab Changed the Oil Business,'
Fortune, XXIX, No. 2 (August, 1971).

'Britain Will Withdraw from the Gulf,' *Middle East Economic Digest,* XIV, No. 15 (December 18, 1970).

'Britain's Energy Pattern,' *Petroleum Press Service,* XXXVII, No. 6 (June, 1970).

Bujra, A.S. 'Urban Elites and Colonialism: The Nationalist Elites of Aden and South Arabia,' *Middle Eastern Studies,* VI, No. 2 (May, 1970).

Burck, Gilbert. 'World Oil: The Game Gets Rough,' *Fortune,* LVII, No. 5 (May, 1958).

'Cheapness with Security,' *Petroleum Press Service,* XXXV, No. 1 (January, 1968).

Cohen, Paul. 'The Erosion of Surface Naval Power,' *Foreign Affairs,* XLIX, No. 2 (January, 1971).

'Consumers Pay the Price,' *Petroleum Press Service,* XXXVIII, No. 4 (April, 1971).

'Control of the Gulf,' *Christian Science Monitor* (December 8, 1971).

Cooley, John K. 'Algerian-Libyan Common Oil Front Faces Solidarity Test Soon?' *Christian Science Monitor* (April 30, 1971).

_____. 'French Take Spill on Algerian Oil,' *Christian Science Monitor* (April 16, 1971).

_____. 'Iraq Charged with Mass Deportation of Iranians,' *Christian Science Monitor* (January 8, 1972).

_____. 'It's Several Tails Wagging Several Dogs in Middle East,' *Christian Science Monitor* (December 4, 1971).

_____. 'New Iraq Accord Nearly Doubles Oil Income,' *Christian*

Science Monitor (June 24, 1971).

_____. 'Soviets Push Interests in Asia,' *Christian Science Monitor* (April 12, 1972).

_____. '10-Power Oil Talks Set Precedent,' *Christian Science Monitor* (January 29, 1971).

'Current Analysis, Oil,' *Standard and Poor's Industry Surveys,* CXXXIX, No. 24, Sec. 1 (June 17, 1961).

'Defense, State Okay Imports of Algerian LNG,' *Oil and Gas Journal,* LXIX, No. 31 (August 2, 1971).

'Demand Will Rise 7.4% This Year,' *World Petroleum,* XLI, No. 9 (September, 1970).
Dold, Paul. 'African Reaction: Power Plays on Black, White Tensions,' *Christian Science Monitor* (November 18, 1970).

Drysdale, John. 'South Asia Wants an Open Door,' *Christian Science Monitor* (November 20, 1970).

DuVal, Dibrell. 'Important Shifts Occur on US Import Scene,' *Oil and Gas Journal,* LXVIII, No. 29 (July 20, 1970).

'European Community,' *Petroleum Press Service,* XXXVIII, No. 2 (February, 1971).

'Europe's National Oil Companies,' *Petroleum Press Service,* XXXVI, No. 11 (November 1969).

Feis, Herbert. 'Oil for Peace or War,' *Foreign Affairs,* XXXII, No. 3 (April 1954).

Gardner, Frank. 'Algerians Post $2.60/BBL Crude Price,' *Oil and Gas Journal,* LXIX, No. 16 (April 16, 1971).

Bibliography

_____. 'A King Speaks,' *Oil and Gas Journal*, LXX, No. 9 (February 26, 1972).

_____. 'International Oil Beset on Every Side,' *Oil and Gas Journal*, LXIX, No. 3 (January 18, 1971).

_____. 'Libyan Oil Agreement Makes Big Waves,' *Oil and Gas Journal*, LXIX, No. 15 (April 12, 1971).

_____. 'OPEC to Ponder New Participation Jolt,' *Oil and Gas Journal*, LXX, No. 11 (March 13, 1972).

_____. 'OPEC Faced with Collective Bargaining,' *Oil and Gas Journal*, LXIX, No. 4 (January 25, 1971).

_____. 'Persian Gulf Oils Bow to Participation,' *Oil and Gas Journal*, LXX, No. 13 (March 27, 1971).

_____. 'Soviets Chortle Over Gas Riches, US Supply Pinch,' *Oil and Gas Journal*, LXVIII, No. 36 (September 7, 1970).

_____. 'World Oil Industry Licks Its Wounds, Plans Ahead,' *Oil and Gas Journal*, LXIX, No. 25 (June 21, 1971).

Grady, Henry F. 'What Went Wrong in Iran?' *Saturday Evening Post*, CCXXIV (January 5, 1952).

'Growing Soviet Economic Stake in Middle East,' *Middle East Economic Digest*, XIV, No. 35 (August 28, 1970).

Hagemann, Erhard. 'Germany Weighs Soviet Crude Offer,' *World Petroleum*, XL, No. 8 (August, 1969).

Hague, Brian C. 'Sabiriyah Raises Kuwait Production,' *World Petroleum*, XLI, No. 11 (November, 1970).

Heller, Charles A. 'American Petroleum Fortress?' *World Petroleum*,

XL, No. 6 (June, 1969).

_____. 'The Strait of Hormuz-Critical in Oil's Future,' *World Petroleum*, XL, No. 11 (October, 1969).

Holden, David. 'The Persian Gulf: After the British Raj,' *Foreign Affairs*, XLIX, No. 4 (July, 1971).

Hoskins, Halford L. 'Changing of the Guard in the Middle East,' *Current History*, LII, No. 306 (February, 1967).

_____. 'Needed: A Strategy for Oil,' *Foreign Affairs*, XXIX, No. 2 (January, 1951).

Hoyt, Monty. 'Oil Output on Verge of Decline,' *Christian Science Monitor* (June 7, 1971).

_____. 'US Burning Its Way Toward Fuel Crisis,' *Christian Science Monitor* (June 3, 1971).

Hudkins, A.H. 'Gas and Oil,' *World Oil*, CLXXI, No. 6, (November, 1970).

Hughes, Edward. 'The Russians Drill Deep in the Middle East,' *Fortune*, LXXVIII, No. 1 (July, 1968).

'Importance Growing as Supplier of Crude Oil,' *World Oil*, CLXXI, No. 3 (August 15, 1970).

'Iran Accord Asked by US and Britain,' *The New York Times* (April 10, 1951).

'Iran Threatens Oil Stoppage,' *Christian Science Monitor* (January 26, 1971).

'It's Official: Foreign Oil Costs More,' *Oil and Gas Journal*, LXVIII, No. 38 (September 21, 1970).

Jablonski, Wanda. 'Master Stroke in Iran,' *Collier's* (January 21, 1955).

Jansen, Godfrey. 'Crisis Potential in the Persian Gulf,' *Christian Science Monitor* (October 29, 1970).

Johansson, Bertram. 'Bad Feeling Over Oil Hikes, Libyan Housecleaning Gets Few Plaudits from Abroad,' *Christian Science Monitor* (May 26, 1971).

Kelly, J.B. 'The British Position in the Persian Gulf,' *World Today*, XX (June, 1964).

Kinney, Gene T. 'Nixon Moves to Aid Foreign Oil Talks,' *Oil and Gas Journal*, LXIX, No. 4 (January 25, 1971).

Knowles, Ruth Sheldon. 'A New Soviet Thrust,' *Mid East: A Middle East and North Africa Review*, IX (December, 1969).

Laqueur, Walter. 'Russia Enters the Middle East,' *Foreign Affairs*, XLVII, No. 2 (January, 1969).

Lee, Christopher D. 'Soviet and Chinese Interest in Southern Arabia,' *Mizan*, XIII, No. 1 (August, 1971).

Levy, Walter J. 'Oil Power,' *Foreign Affairs*, XLIX, No. 4 (July, 1971).

'Libya Pressuring Concessionaires to Boost Drilling,' *Oil and Gas Journal*, LXVIII, No. 23 (June 8, 1970).

'Lincoln Pleads for US Energy Plan,' *Oil and Gas Journal*, LXIX, No. 4 (January 25, 1971).

Lockhart, Laurence. 'The Causes of the Anglo-Persian Oil Dispute,' *Journal of Royal Central Asian Society*, XL (April, 1953).

Luce, William. 'Britain in the Persian Gulf: Mistaken Timing Over Aden,' *Round Table*, LVII (July, 1967).

May, John Allan. 'British and US Strategy: Thwarting Soviet Influence,' *Christian Science Monitor* (November 17, 1970).

_____. 'Russians in the Indian Ocean,' *Christian Science Monitor* (November 17, 1970).

McKinsey, Philip W. 'US Trustbusters Scan Oil-Industry Operations,' *Christian Science Monitor* (January 6, 1971).

Millar, T.B. 'Soviet Policies South and East of Suez,' *Foreign Affairs*, XLIX, No. 1 (October, 1970).

Mitchell, Donald W. 'The Soviet Naval Challenge,' *Orbis*, XIV, No. 1 (Spring, 1970).

'Money Woes Crimping Japan's Ability to meet Skying Needs,' *Oil and Gas Journal*, LXVIII, No. 31 (August 3, 1971).

'More Domestic Oil Vital for US Health,' *Oil and Gas Journal*, LXIX, No. 7 (February 15, 1971).

Morison, David. 'Soviet Involvement in the Middle East: The New Strategy,' *Mizan*, XI, No. 5 (September-October, 1969).

'North American Reserves–1969,' *Petroleum Press Service*, XXXVIII, No. 5 (May, 1970).

'North Rumaila Oil for Russia,' *Petroleum Press Service*, XXXVI, No. 8 (August, 1969).

O'Hanlon, Thomas. 'Mobil Oil "Squarely in the Middle East",' *Fortune*, LXXVI, No. 3 (September, 1967).

'Oil Firms Team Up on Mideast,' *Christian Science Monitor* (January 20, 1971).

'Oil is Taking Over the Energy Business,' *Business Week*, No. 2149 (November 7, 1970).

'Oil Production Still Restricted,' *Middle East Economic Digest*, XIV, No. 25 (June 19, 1970).

'Oil Unit Demands A Say,' *Christian Science Monitor* (October 12, 1971).

'OPEC Gets to Essentials,' *The Economist*, CCXVI (August 28, 1965).

'OPEC Seeks a Plan,' *Petroleum Press Service*, XXXVII, No. 9 (September, 1970).

'OPEC Stepping up State-Participation Drive,' *Oil and Gas Journal*, LXIX, No. 31 (August 2, 1971).

'OPEC Strongly Backs All Arab Claims,' *Oil and Gas Journal*, LXVIII, No. 32 (August 10, 1970).

'OPEC Working for Even Bigger Share,' *Oil and Gas Journal*, LXIX, No. 1 (January 4, 1971).

Page, Stephen. 'Moscow and the Persian Gulf Countries, 1967-1970,' *Mizan*, XIII, No. 2 (October, 1971).

'Persian Gulf Crude Prices Jump Again,' *Oil and Gas Journal*, LXIX, No. 24 (June 14, 1971).

'Persian Gulf Vacuums,' *Christian Science Monitor* (March 4, 1971).

Pogue, Joseph E. 'Must an Oil War Follow This War?' *Atlantic Monthly*, CLXXIII, No. 3 (March, 1944).

'Rising Demand for Oil Met in 1970,' *Middle East Economic Digest,* XV, No. 2 (January 8, 1971).

Roosevelt, Edith Kermit. 'Oil, Arabs and Communism,' *America,* CXIX (September 21, 1968).

'Russia and Arab Oil,' *Petroleum Press Service,* XXXV, No. 2 (February, 1968).

'Russia Drives East of Suez,' *Newsweek* (January 18, 1971).

'Russian Reserves Are Inadequate,' *Petroleum Press Service,* XXXVI, No. 4 (April, 1969).

Scott, Robert W. 'Petroleum in the Year 2000,' *World Oil,* CLXXI, No. 3 (August 15, 1970).

'Shah: Eliminate Oil Production System,' *Middle East Economic Digest,* XIV, No. 25 (June 19, 1970).

Sharp, Roger R. 'America's Stake in World Petroleum,' *Harvard Business Review,* XXVIII, No. 5 (September, 1950).

'Slow Decline of Soviet Exports,' *Petroleum Press Service,* XXXVII, No. 5 (May, 1970).

'Soviet Bloc is Hardening,' *Petroleum Press Service,* XXXV, No. 11 (November, 1968).

'Soviet Drilling Costs up as Technology Lags Behind,' *World Petroleum,* XL, No. 7 (July, 1969).

'Soviet Export Rise Halted,' *Petroleum Press Service* XXXV, No. 6 (June, 1968).

'Soviet Interest in Middle East Oil,' *Mizan,* XIII, No. 1 (August, 1971).

'Soviet Oil in the "Seventies",' *Petroleum Press Service*, XXXVII, No. 1 (January 1970).

'Soviet Oil Minister Sees 10 Million B/D Flow by 1975,' *Oil and Gas Journal*, LXVII, No. 34 (August 24, 1970).

'Soviet Oil Strategy,' *Middle East Economic Digest*, XIV, No. 47 (November 20, 1970).

'Soviets Face Expansion Slowdown in '71,' *Oil and Gas Journal*, LXIX, No. 1 (January 4, 1971).

Standish, J.F. 'Britain in the Persian Gulf,' *Contemporary Review*, CCXI, No. 1222 (November, 1967).

'Stripper-well, Survey Shows Production Edging Down Slightly,' *Oil and Gas Journal*, LXIX, No. 4 (January 25, 1971).

Strout, Richard L. 'What Oil Quotas Cost US Families,' *Christian Science Monitor* (February 2, 1972).

'A Successful Bargaining Scheme,' *World Petroleum*, XLI, No. 10 (October, 1970).

'Suez Versus Supertankers,' *Middle East Economic Digest*, XV, No. 7 (February 12, 1971).

Sullivan, Robert R. 'The Architecture of Western Security in the Persian Gulf,' *Orbis*, XIV, No. 1 (Spring, 1970).

'Task: Find 250-450 Billion BBL,' *Oil and Gas Journal*, LXVIII, No. 26 (June 29, 1970).

Thoman, Roy E. 'The Persian Gulf Region,' *Current History*, LX, No. 353 (January, 1971).

Todd, William F. 'The Impact of Oil on Middle East Economies,'
    World Petroleum, XL, No. 1 (January, 1969).

'Towards Bilateral Arrangements,' *Petroleum Press Service*, XXXVIII,
    No. 5 (June, 1971).

'US Asks Persia to Modify Oil Policy,' *The Times* (London) (June 28,
    1951).

'US Bids Iran Resist Threats as Debate on Soviet Oil Nears,' *The
    New York Times* (September 12, 1947).

'US Dependence on Oil Imports,' *Petroleum Press Service*, XXXVIII,
    No. 5 (May, 1971).

'US Energy Shortage Said Worst of Century,' *Oil and Gas Journal*,
    LXVIII, No. 26 (June 29, 1970).

'US Investment in Arab World Tops $2000M,' *Middle East Economic
    Digest*, XIV, No. 34 (August 21, 1970).

'US Oil Supply Losing Race to Demand,' *Oil and Gas Journal*, LXIX,
    No. 3 (January 18, 1971).

'US Policy on Persian Oil,' *The Times* (London) (April 10, 1951).

Ushijima, Toshiaki. 'Japan's Vigorous Oil Search Widens,' *World
    Petroleum*, XLVXXI, No. 12 (December, 1970), Table 1.
Walsh, Maximilian. 'Australia Expands Defense Role to West,'
    Christian Science Monitor (November 20, 1970).

Watt, D.C. 'Britain and the Future of the Persian Gulf States,' *World
    Today*, XX (November, 1964).

'Western Europe: Inland Sales in 1970,' *Petroleum Press Service*,
    XXXVIII, No. 1 (January, 1971).

'What Project Rulison Can Mean to US Domestic Energy Supply,' *World Oil*, CLXXI, No. 6 (November, 1970).

'When Oil Flows East,' *The Economist*, CCXXXIV, No. 6594 (January 10, 1970).

'Who Controls Electric Energy Controls Us,' *Christian Science Monitor* (August 7, 1971).

'Why the Search Goes On,' *Petroleum Press Service*, XXXVII, No. 6 (June, 1970).

Zoppo, Ciro. 'Soviet Ships in the Mediterranean and the U.S. Soviet Confrontation in the Middle East,' *Orbis*, XIV, No. 1 (Spring, 1970).

Index

Note: numbers in **bold** refer to tables and numbers in *italics* refer to figures.

Index